PROCEEDINGS
OF THE
TWENTY-FIFTH
ANNUAL MEETING
5–6 APRIL, 1989

RADIATION PROTECTION TODAY—THE NCRP AT SIXTY YEARS

Warren K. Sinclair, Editor

RA569
N355
1989

**As Presented at the
Vista International Hotel,
Washington, D.C.**

Issued April 1, 1990

National Council on Radiation Protection and Measurements
7910 WOODMONT AVENUE / Bethesda, MD 20814

The material included in these proceedings has not been reviewed and represents the views of the authors and not necessarily those of the NCRP:

Copyright © National Council on Radiation
Protection and Measurements 1989

All rights reserved. This publication is protected by copyright. No part of this publication may be reproduced in any form or by any means, including photocopying or utilized by any information storage and retrieval system without written permission from the copyright owner, except for brief quotation in critical articles or reviews.

Preface

The 25th Annual Meeting of the National Council on Radiation Protection and Measurements was a special anniversary year for the NCRP. Not only was it the 25th meeting of the Council but it was also the 60th anniversary of the forerunner organizations of the Council, initiated first in 1929.

In honor of the occasion the NCRP adopted as its theme for the meeting "The Status of Radiation Protection Today—The NCRP at 60 Years," with the aim of establishing how we came to where we are and where we might be going in the future. The NCRP also proposed to repeat this type of status program each decade.

Accordingly, the program began with a history and perspective on the development of radiation protection standards and ended with a discourse on radiation protection policy and its implications for the future. In between, the program dealt with the scientific basis of radiation protection (risk estimation, somatic and genetic, and experimental radiobiology), perception, and experience (in power plants, in medicine, in occupational and environmental exposures and with accidents). It also included experience and progress with nonionizing radiation, (electromagnetic, measurement and ultrasound) as well as with ionizing radiation measurement, dosimetry and the status and trends in radiation regulation.

The 13th L.S. Taylor Lecture was presented by Arthur C. Upton on "Radiobiology and Radiation Protection: The Past Century and Future Prospects." The speaker was introduced by Naomi H. Harley.

The Council is indebted to the authors who participated in the program and provided the papers published here. The Council acknowledges its debt to the James Picker Foundation for past support of the Taylor Lecture series and NCRP Publications.

On behalf of the Council I want to express my appreciation for the fine work of my Co-Chairman, Dade Moeller, and the other members of the Program Committee.

Dade W. Moeller, Co-chairman
Harvard University
Boston, Massachusetts

Warren K. Sinclair, Co-chairman
National Council on Radiation Protection and Measurements
Bethesda, Maryland

Merril Eisenbud
University of North Carolina
Chapel Hill, North Carolina

Charles B. Meinhold
Brookhaven National
　Laboratory
Upton, New York

John E. Till
Radiological Assessment
　Corporation
Neeses, South Carolina

Arthur C. Upton
New York University Medical
　Center
New York, New York

George M. Wilkening
AT&T Bell Laboratories
Murray Hill, New Jersey

I served as Editor of these Proceedings and I wish to thank Laura Martin who provided excellent editorial assistance.

Bethesda
July 31, 1989

Warren K. Sinclair
President

Contents

Preface	iii
Opening Remarks	1
Warren K. Sinclair, *President*	
Scientific Session: Historical Perspective	3
Merril Eisenbud, *Chairman*	
History and Perspective on the Development of Radiation Protection Standards	5
Dade W. Moeller	
Scientific Session: Scientific Basis	25
Gilbert Beebe, *Chairman*	
The Status of Somatic Risk Estimation	27
William J. Schull	
Experimental Radiobiology and Radiation Protection Studies	45
R.J. Michael Fry	
Genetic Risk Estimates: Of Mice and Men	61
Seymour Abrahamson	
Scientific Session: Perception	71
Leonard A. Sagan, *Chairman*	
Perception of Radiation Risks	73
Paul Slovic	
Scientific Session: Radiation Protection Experience	99
Charles B. Meinhold, *Chairman*	
Occupational Exposures and Practices in Nuclear Power Plants	101
John W. Baum	
Status and Trends in Radiation Protection of Medical Workers	124
William R. Hendee and Marc Edwards	
Reconstructing Historical Exposures to the Public from Environmental Sources	146
John E. Till	
Medical Exposures	161
Fred A. Mettler	
An Historical Perspective of Human Involvement in Radiation Accidents	171
C.C. Lushbaugh, Shirley A. Fry, A. Sipe and Robert C. Ricks	

v

Opening Remarks to the Taylor Lecture 191
   Warren K. Sinclair
Introduction of the Taylor Lecturer 193
   Naomi H. Harley
Taylor Lecture: Radiobiology and Radiation Protection: The
Past Century and Prospects for the Future 199
   Arthur C. Upton
Scientific Session: Nonionizing Radiation 233
   S. James Adelstein, *Chairman*
Protection Against Nonionizing Electromagnetic Radiation,
An Evolutionary Process 235
   George M. Wilkening
The State of the Art in Measuring Electromagnetic Fields for
Hazard Assessment 251
   Richard A. Tell
Protection in Medical Ultrasound: Modest Progress in a
Low-Risk Field 262
   Paul L. Carson and J. Brian Fowlkes
Scientific Session: Measurement and Dosimetry 279
   Randall S. Caswell, *Chairman*
Radiation Protection Measurement Science: A Long Road
Traversed but a Long Way to Go 281
   Gail de Planque, Wayne N. Lowder and Harold L. Beck
Radiation Dosimetry: Past and Present 299
   Kenneth R. Kase
Scientific Session: Regulation of Ionizing and Nonionizing
Radiation and Implications for the Future 331
   Dade W. Moeller, *Chairman*
Development and Current Trends Regarding Regulation of
Radiation Protection 333
   Manning Muntzing
Radiation Protection Policy: Implications for the Future 343
   Warren K. Sinclair
NCRP Publications 359

TWENTY-FIFTH ANNUAL MEETING OF THE NATIONAL
COUNCIL ON RADIATION PROTECTION AND
MEASUREMENTS

Radiation Protection Today—The NCRP at Sixty Years

Opening Remarks

Warren K. Sinclair
President, NCRP

It is a pleasure to welcome you all to this 25th anniversary meeting of the National Council on Radiation Protection and Measurements. This occasion is also the 60th anniversary of the founding of the predecessor organizations of the NCRP, the original Advisory Committee on X-Ray and Radium Protection, in 1929, later (in 1946) the National Committee on Radiation Protection and finally in 1964 the National Council on Radiation Protection and Measurements. We welcome especially visitors to Washington. The cherry blossoms were a little early this year but they are still in full bloom and there is still time, preferably today, to see them in all their glory.

May I draw your attention to the pamphlet labelled NCRP News which was available at the registration desk. It describes some of the recent activities of the Council, notably in the form of published reports. I'd like to draw your attention not only to the list of reports published last year but also to two even more recent reports #96, on Comparative Carcinogenicity of Ionizing Radiation and Chemicals which covers new ground for the NCRP and report #99 on Quality Assurance for Diagnostic Imaging Equipment, an important new work in radiological protection.

Now let me turn to the program for this significant anniversary meeting. Some years ago it was suggested by my Co-chairman, Dade Moeller, that the NCRP do a "status of radiation protection" program as sort of

1

"show and tell" on where we stand, what we have all done well in radiation protection and what we have done less well and need to improve. The Board thought this was an excellent idea and considered also that we should hold it on our anniversary year and that we should repeat it at decade intervals. That is now the plan. This is the first of these "show and tell" programs, consequently, it has to consider the whole of our past experience with radiation rather than merely the past decade as future programs will do. Dade's original idea has been embellished and perhaps also obfuscated by his Co-chairman, by the other members of the program committee and by the Board itself. As a result, we have the program we look forward to today.

I will not detail it to you, but it will suffice to point out the broad outlines. First, Dade Moeller will review our past and new perspectives on radiation protection standards. Then we will proceed to describe epidemiological and laboratory findings on which our risk estimates of low dose stochastic effects depend. We will have one talk on perceptions of those risks by the public who are exposed to them. This afternoon we will turn to radiation protection experience in occupational, nuclear workers and medical; public, environmental, and medical exposures and finally a discussion of accidents as part of our overall experience. The Taylor Lecture by Arthur Upton will deal with the appropriate subject of the relationship between radiation biology and radiation protection and will conclude today's presentations.

Tomorrow we will have a full session on nonionizing radiation and trace some of our experience there in measurement and in control of electromagnetic and ultrasound radiations. This will be followed by a return to ionizing radiation, its measurement and dosimetry and how these have changed over the years. The last session will deal with some of our experience and trends in regulation and finally I will try to put radiation protection and the role of NCRP in the future in perspective.

We have made some changes in the structure of the program for this anniversary year, the business meeting was held early and we have not included brief committee reports this year to make time available for the anniversary program. We expect to include the committees again in the future.

The program committee has provided a distinguished list of speakers for us, so let us get on with it by turning the meeting over now to our first chairman and distinguished honorary member, Merril Eisenbud.

Scientific Session

Historical Perspective

Merril Eisenbud
Chairman

History and Perspective on the Development of Radiation Protection Standards

Dade W. Moeller
Harvard School of Public Health
Boston, MA

Abstract

From a beginning in which the goal was to prevent acute effects in occupational settings, the development of radiation protection standards has progressed to an age where they are being set not only to limit long-range effects but also to control doses from sources of natural origin and to protect susceptible population groups. Playing a major role in these developments are quantitative evaluations of the associated risks to health. Under the ALARA criterion, facilities are being designed and operated so that the average doses to workers are only a small fraction of the limits. Recent efforts include the development of exemption levels for sources considered to be "Below Regulatory Concern," and proposals for dose rates considered to be *de minimis*. Many of these concepts would appear to have useful applications in other fields of occupational and environmental health.

"All the world's a stage
And all the men and women merely players.
They have their exits and their entrances;
And one man in his time plays many parts,
His acts being seven ages" (1).

5

From a beginning in which the goal was to prevent acute effects in occupational settings, the development of radiation protection standards has progressed to an age where they are being set not only to limit long-range effects but also to control doses from sources of natural origin and to protect susceptible population groups. Just as with men and women, the history or life of the development of radiation protection standards appears to have had seven ages. Hopefully, in the case of radiation protection standards, the current age is one of professional maturity, not one in which we are undergoing "second childishness, and mere oblivion, sans teeth, sans eyes, sans taste, sans everything" (1).

Age Number One: Avoidance of Acute Effects (1900–1930)

Shortly after the discovery of x-rays in 1895 and of naturally occurring radioactive materials in 1896, reports of radiation injury began to appear in the published literature (2). Recognizing the need for protection, dose limits were informally recommended with the primary initial concern being to avoid direct physical symptoms. These limits were based on the work of Mutscheller (3) and led to a guideline, first of one-tenth an erythema dose per year and later to one one-hundredth of an erythema dose per month. Assuming that an erythema dose was about 600 R, this was equivalent to a dose limit of about 60 R (0.6 Sv) per year.

As early as 1902, however, it had been suggested that radiation exposures might result in delayed effects, such as the development of cancer (4). This was subsequently confirmed for external sources and, between 1925 and 1930, it was also confirmed in the case of exposures from internally deposited radionuclides when bone cancers were reported among radium dial painters (2). With the publication by H. J. Muller in 1927 of the results of his experiments with Drosophila, concern began to be expressed regarding the possibility of genetic effects from radiation exposures in humans (5). These events, in turn, led to the second age in the development of radiation protection standards.

Age Number Two: Concern for Chronic Effects (1930–1950)

During Age Number One, the subject of dose limits for radiation workers was reviewed informally, with gradual reductions being recommended in the limits as greater technological control became feasible and as additional knowledge of potential long term effects was gained. This approach changed with the formation in 1928 of the International X-ray and Radium Protection Committee (known today as the International Commission on Radiological Protection) and of the U. S. Advisory

Committee on X-Ray and Radium Protection (known today as the National Council on Radiation Protection and Measurements). In 1934, the first formal "standards" for external radiation protection were proposed by the U. S. Advisory Committee (6). Even though this report was primarily directed to radium, it contained the statement that "The safe general radiation to the whole body is taken as 1/10 R (1 mSv) per day for hard x rays.." Added to these recommendations in 1941 were proposed limits on the intake of radium (7). With the development of nuclear weapons in World War II and the increasing use of all types of radiation sources, the development of radiation protection standards entered a new age.

Age Number Three: Concern for Genetic Effects (1950–1960)

One of the initial observations (later proven not to be accurate) resulting from the epidemiology studies in Japan was an apparent change in the ratio of males to females among infants born to the survivors of the atomic bombings (8). This observation, coupled with the general acceptance of the studies by Muller, let to increasing concern relative to the potential genetic effects of radiation. In fact, this concern dominated the bases for the development of radiation protection standards from the end of World War II until about 1960 and led to the first consideration of recommendations for dose limits to the public. This was exemplified by the report of the NAS-NRC Committee on the Biological Effects of Atomic Radiation (BEAR) issued in 1956 which called for (a) a reduction in the existing occupational dose limits, (b) increased concern for genetic risks, and (c) establishment of dose limits for the general public (9). A major conclusion of the BEAR study was that the principal limitation on radiation exposures for members of the public was the associated genetic risk. It is of interest to note that the Medical Research Council in the United Kingdom issued at the same time a similar report which presented essentially the same conclusions (10). As a result of these and other considerations, the National Council on Radiation Protection and Measurements (NCRP) reduced in 1958 its recommendation for average annual occupational whole body dose limits from 0.15 to 0.05 sievert (15 to 5 rem) (11).

During the period from 1957 to 1959, the NCRP began to address the issue of exposures to the general public. At the same time, the Council adopted the principle that "radiation or radioactive material outside a controlled area attributable to normal operations within the controlled area, shall be such that it is improbable that any individual will receive a dose of more than 0.5 rem (5 mSv) in any year from external radiation." It was further stated that "the maximum permissible body-burden of radionuclides in persons outside of the controlled area and attributable to operations within the controlled area shall not exceed 1/10 of that for radiation workers" (12).

In Report No. 17, published in 1954, (13) the NCRP acknowledged the problem of genetic risks and expressed increasing concern about the potential latent effects of radiation exposures, particularly leukemia. In its Publication 1, published in 1959, (14) the International Commission on Radiological Protection (ICRP) made similar observations. This concern for leukemia was to serve as a primary basis for the establishment of radiation protection standards over the next decade.

Age Number Four: Concern for Somatic Effects (Primarily Leukemia) (1960–1970)

With the establishment of the Federal Radiation Council (FRC) in 1959, additional attention was focused on dose limits for the general public. In 1960, the FRC issued its first recommendations (15). With respect to occupational exposures, its recommended guidelines were essentially the same as those of the NCRP and the ICRP. In making recommendations for the general public, the FRC proposed that the maximum whole body dose rate to any individual in the population not exceed 5 mSv (0.5 rem) per year. Although not explicit at the time, it is to be presumed that this maximally exposed individual was located within what today is commonly referred to as the "critical group." In order to facilitate interpretation of data on average exposures, the FRC assumed that doses to "the majority of individuals do not vary from the average by a factor greater than three." Based on this assumption, the FRC concluded that the average dose rate for a population group would not exceed 1.7 mSv (170 mrem) per year.

In Publication 1, the ICRP recommended that "the genetic dose to the whole population from all sources additional to natural background shall not exceed 5 rem (50 mSv) plus the lowest practicable contribution from medical exposures" (14). Because the ICRP considered the genetically significant dose to be that accumulated within the first 30 years of life, this recommendation translated into an average dose rate of 1.7 mSv (0.17 rem) per year per individual within the general population, exactly the same limit as recommended by the FRC.

Commencing during the preceding age and continuing through this age, standards recommending groups were also interested in the avoidance of other types of latent effects. Each of the various advisory and/or regulatory groups, however, used different words or phrases to express their goals relative to the control of these other types of effects. For example, the key phrase used by the NCRP was to avoid "observable injury during the lifetime of the exposed individual" (13). The goal of the FRC, in contrast, was to avoid "undue hazard," (15) while the goal of the U.S. Nuclear Regulatory Commission was to provide "a substantial margin of safety" (16). Although the goals were common, the methods of expressing them helped each advisory body or regulatory group maintain its individuality.

Age Number Five: Concern for Somatic Effects (Primarily Solid Tumors) (1970–the Present)

In 1960, the BEAR Committee issued a report (8) in which it acknowledged that the risks from genetic effects had been overestimated in its earlier report. This was followed by the 1961 report of the United Nations Scientific Committee on the Effects of Atomic Radiation (UNSCEAR) (17) which concurred that genetic risks had previously been overestimated. In addition, the UNSCEAR report noted the sensitivity of the fetus to radiation which, in turn, could lead to childhood leukemia. This was followed in 1966 by ICRP Publication 8 in which the Commission concluded that the major consequences of whole-body exposures would be an increased incidence of acute leukemia and chronic myeloid leukemia (18).

Subsequent to these developments, however, detailed analyses of the increasing amount of data being developed through human epidemiological studies showed that solid tumors, rather than leukemia, were the governing somatic risks. This was highlighted by Report No. 39, published by the NCRP in 1971, (19) by the UNSCEAR report of 1977, (20) and by the 1980 report of the NAS-NRC Committee on the Biological Effects of Ionizing Radiation (BEIR), (21) all of which confirmed the importance of solid tumors.

Age Number Six: Application of a Risk-Based Approach (1980–the Present)

One of the earliest reports that provided estimates of the somatic risks of radiation exposures was the paper by Lewis, published in 1957, on the risk of leukemia due to fallout from atmospheric tests of nuclear weapons (22). Subsequent information was provided in the 1972 report of the NAS-NRC BEIR Committee (23) and by the 1972 and 1977 UNSCEAR reports (24, 20). These latter reports provided detailed quantitative estimates of both the genetic and somatic (cancer) risks resulting from population exposures to ionizing radiations. One of the principal conclusions of the groups developing these reports was that it should not be taken for granted that genetic risks from exposures of populations to dose rates near natural background levels are of greater importance than the somatic risks. This provided additional support to the earlier analyses cited above.

During the last half of the 1970s, further progress was made in applying the risk approach in setting dose limits. This was highlighted by the issuance in 1977 of ICRP Publications 26 and 27 which formally introduced a methodology for application of the risk approach in setting dose limits and made a major contribution to the establishment of standards based on risk/benefit assessments (25, 26).

In applying the risk-based approach to the establishment of radiation protection standards for radiation workers, the basic philosophical criterion has been that the average incremental risk to radiation workers due to their radiation exposures should be no greater than the average incremental risk of death from traumatic injuries to workers employed in "safe" industries. A significant outgrowth of the application of this approach was the development of the concept of the effective dose equivalent which enabled doses to single body organs to be expressed in terms of an equivalent dose to the whole body. This, in turn, led to a mathematical system in which doses from internal and external sources could be summed.

One of the many benefits of this system has been to demonstrate even more conclusively the manner in which radiation exposures from natural background dominate the dose equivalents being received by the world's population today. Using the effective dose equivalent approach, for example, radon and its airborne decay products inside buildings have been shown to be the source of over half of the total radiation dose to the U.S. public today (27). Another major benefit of the risk-based approach has been the wide recognition that efforts should be made to maintain average dose rates to radiation workers at a level only a small fraction (about $1/10$) of the dose limit. This has provided an increasing incentive for the implementation of the ALARA criterion as a routine part of all radiation protection programs. It has also provided additional incentives to organizations, such as the NCRP, to give additional consideration to what should be considered to be permissible dose rates in the future. Summarized in Figure 1 are the occupational dose limits recommended by the NCRP over the past 50 years, including their latest recommendations as published in Report No. 91 (28). Also indicated on the Figure is the limit based on the recommendation of Mutscheller in 1925, (3) plus what is generally considered today as an acceptable average dose

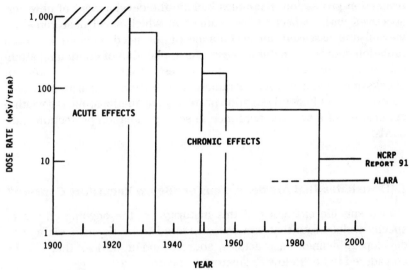

Fig. 1. Occupational Dose Limits

rate for a group of workers, assuming the proper application of the ALARA criterion. This latter vaue is shown at 10% of the dose rate limit.

Although use of the risk-based approach has proven to be exceedingly useful, it has not been without its difficulties. These include questions on whether the risk should be based on fatalities, only, or if the impacts of morbidity should be included; whether genetic effects should be considered for only the next several, or for all future, generations; and whether the evaluations that are being applied to workers can be similarly used for setting protection standards for members of the public. Also to be considered is the fact that the radiation worker is exposed to the potentiality of death from traumatic injuries similar to those in non-radiation industries. Finally, although this approach has provided a mechanism for expressing the doses to single organs in terms of the equivalent whole body dose, yet to be developed is a comparable system that will enable partial body doses from external sources to be expressed in terms of an equivalent whole body dose. Nonetheless, the risk-based approach has proven to be a sound basis on which to proceed and it will undoubtedly serve as a foundation for the continuing development of radiation protection standards for some time to come.

Age Number Seven: The Approach to Professional Maturity (1990– the Future)

Beginning in the late 1980s, and hopefully setting the stage for the 1990s, there have been many indications of the maturing of the radiation protection profession. Examples include the development of new perspectives with respect to the manner in which small dose rates are viewed and assessed, an acceptance of the need to control certain radiation sources of natural origin, the application of continuing attention to special population groups, the establishment of a system of checks and balances in setting radiation protection standards, improved coordination of Federal radiation protection regulations, and the routine application of a systems approach in setting radiation protection standards.

1. Dose Rates That Are *de minimis* or "Below Regulatory Concern"

Two specific examples of this maturity are the ongoing efforts to specify a dose rate that can be considered to be *de minimis* and to develop guidelines on practices, sources and/or devices that can be considered to be "Below Regulatory Concern."

a. The Concept of a *de minimis* Dose

For the past decade, efforts have been underway to specify a dose rate to individual members of the public that can be considered to be *de minimis*. Although many public health and radiation protection experts have suggested that a dose rate in the range of 10 µSv (1 mrem) per year or less might be so considered, progress in obtaining approval of such a concept by regulatory agencies has been slow. One significant advancement in this regard was the recommendation by the NCRP in its Report No. 91 (28) that dose rates below 10 µSv (1 mrem) per year not be included in computations of the collective dose equivalent from radiation sources. One of the basic premises for this recommendation was that the postulated increase in the number of cancers resulting from such a dose rate, even if applied to a large population group, would not be sufficient to demand any increase in medical or other types of community services.

b. The Concept of "Below Regulatory Concern"

During the mid-1980s, both the U.S. Environmental Protection Agency (EPA) and the U.S. Nuclear Regulatory Commission (NRC) initiated studies to determine whether certain sources, practices and/or devices that have low associated dose rates might, after careful review and evaluation, be exempted from normal regulatory control. Similar work has been underway within the International Atomic Energy Agency (29).

Examples of items that would be included within the scope of activities being considered by the NRC are low level radioactive wastes, equipment from decommissioned facilities that is being considered for release for public use, and consumer products that produce radiation and/or involve radioactive materials. Efforts of the EPA have been almost exclusively directed to low level wastes.

Factors that must be taken into consideration in making such decisions include dose rates for members of the public (both individually and collectively) that can be considered to be acceptable, how to limit the number of exempted items that impact on a single individual member of the public (particularly when there are several Federal agencies that may be granting such exemptions), the degree to which a given item, source, or device must be socially acceptable, how to define the exposure scenarios (both for routine use and possible accidents) to be considered, and the validation of the pathway models to be used in estimating the potential doses. Although it is too early to determine exactly what criteria, if any, will be approved, indications are that items involving dose rates to individual members of the public in the range of a few tens of microsieverts (a few mrem) per year may be exempted. In order to place a limitation on the number of people affected, particularly by items contributing larger dose rates, it has been suggested that a sliding scale be used. Under this approach, the distribution of exempted items causing higher dose rates to individual members of the public would be restricted in numbers through the imposition of lower collective dose limits. A schematic illustration of these concepts and approaches is shown in Figure 2.

In its initial proposal, the EPA has suggested permitting low level radioactive wastes that involve dose rates no more than 4 mrem (40 µSv) per year to be disposed of in municipal sanitary landfills. The overall concept is an excellent one, it could result in economic savings of millions of dollars per year (permitting these funds to be spent on environmental problems of a more urgent nature), and its application could be extremely beneficial in helping the public gain a better understanding of the health effects of ionizing radiation.

2. Control of Sources of Natural Origin

For years radiation protection standards were developed on the basis that exposures resulting from medical practices and from sources of natural origin were not subject to control. Another example of the

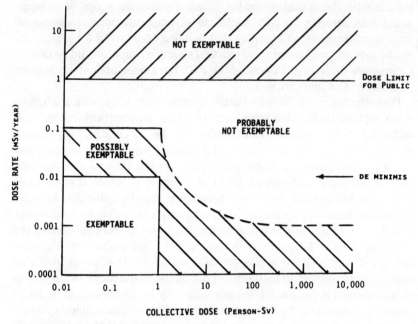

Fig. 2. Approaches for Regulatory Exemptions

maturity of the profession is the acknowledgement that certain sources of natural origin should be subject to control.

A prime example of this is the developing concern about the previously cited exposures to radon inside buildings. Today major efforts are underway to educate the public about this source and to assist them in initiating measures for its control.

3. Consideration of Special Population Groups

Another example of the maturing of the radiation protection profession is the attention that is being directed to special population groups. This is exemplified by the dose limits for the fetus recommended by the NCRP in its Report No. 91, (28) and the considerations being given by the NRC to this matter, as well as to exposures to population groups, in the revisions to Title 10, Part 20, of the Code of Federal Regulations (16). While the NRC recognizes that the re-evaluations of the dose versus health effects data resulting from the epidemiological studies of the Japanese survivors of the World War II bombings may necessitate changes in the dose limits being recommended in their revised standards, they have concluded that any changes necessary can be made after the newer data are confirmed.

4. The Establishment of Checks and Balances in the Development of Radiation Protection Standards

In the early days of the nuclear energy industry, Federal standards for radiation protection were frequently based almost entirely on the recommendations of independent organizations such as the NCRP. As the industry grew and the use of radioactive materials and radiation generating machines became more commonplace, Federal agencies began to develop inhouse radiation protection expertise. As a result, it is more common today for such agencies to use the NCRP recommendations as guidelines rather than as specifications. Prime examples of this approach are the guidance provided by the FRC, beginning in 1960 (15); Title 40, Part 190, of the Code of Federal Regulations (40 CFR 190) through which EPA established standards for the regulation of environmental releases from nuclear power operations (30); 40 CFR 191 in which this same Agency set standards for the disposal of spent nuclear fuel, high level radioactive and transuranic wastes (31); and 10 CFR 50, Appendix I, in which the NRC established limits on releases of radioactive materials from nuclear power plants (32).

5. Improved Coordination of Federal Radiation Protection Regulations

Another mark of maturity in the development of radiation protection standards is the recognition by the various Federal agencies involved of the need for careful coordination of their activities. One example of the benefits of such coordination is the development and publication of the "Radiation Protection Guidance to Federal Agencies for Occupational Exposure," signed by the President in January, 1987 (33). Another example of this type of effort is the work of the Committee on Interagency Radiation Research and Policy Coordination which recently, for example, prepared "A Compendium of Major U.S. Radiation Protection Standards and Guides: Legal and Technical Facts" (34). Included in this report was an enumeration of a set of reforms to correct some of the existing problems. These reforms include recognition that U.S. radiation protection standards are numerous and complex, that they deal principally with control activities that make relatively small contributions to reductions in the overall population dose, and that they do not follow a common rationale in seeking to achieve their public health objectives.

Also reflecting this maturity are the steps that have been taken to assure that members of the public have opportunities to comment on Federal Standards under development. With the promulgation of the Government in the Sunshine Act, essentially all meetings in which such standards are being formulated are conducted in sessions open to the public.

In discussing the problem of coordinating Federal activities related to radiation protection regulations, one must recognize the increasing role that the U.S. Congress is exercising in such activities. One recent example was the requirement specified in the Radon Pollution Control Act of 1988 (35) in which the Congress stipulated that it was their goal that radon concentrations inside buildings should be no higher than ambient concentrations observed outdoors. Whether such an objective is attainable is open to question. However, it is interesting to note that, in taking this action, the Congress has indicated a desire to make the protection standards for this radiation source comparable to those for artificial sources. In contrast, the NCRP and the ICRP have recommended standards for indoor radon that result in dose rates that have associated risks far in excess of those permitted for artificial sources. This is illustrated in Figure 3 (36) which compares the lifetime risks associated with standards that have been established by various Federal agencies and/or recommended by the NCRP for a variety of radiation sources. While the approach of the Congress may represent consistency, that of the NCRP and the ICRP represents a recognition of the practicalities of this situation. The processes used in the resolution of these competing views will be interesting to follow.

6. Application of A Systems Approach

Radiation protection standards for workers and for members of the public are closely coordinated and follow a systematic pattern. One of the basic premises in the establishment of such standards is that dose limits for members of the public should be less than those for workers. Under the current approach, short term annual dose limits and associated risks for members of the public are set at 10%, with long term limits being set at 2%, of the dose limits and associated risks considered acceptable for workers. Average population dose rates are expected to be well below this value, with dose rates considered Below Regulatory

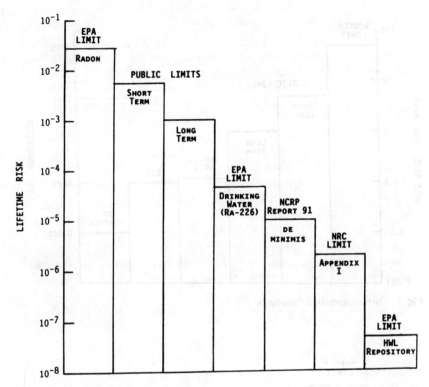

Fig. 3. Comparison of Radiation Standards

Concern being less, and those considered *de minimis* to be in the 10 µSv (1 mrem) per year rate or less (Figure 4). Another important concept is that people younger than the age of 18 should not be employed in radiation work and exposed at occupational dose rates. In addition, through use of the collective dose concept, a mechanism has been developed for estimating and comparing the societal impact of individual practices involving radiation exposures.

It would appear that many of these approaches would be equally applicable in other fields of occupational and environmental health. For example, limits on airborne concentrations for non-radioactive chemical contaminants in the ambient environment (or inside the home) might be set at a specified fraction of those for occupational settings. Some of the considerations in this type of an approach are depicted in Figure 5. In a like manner, the societal impact of toxic chemicals might be estimated on the basis of the number of people exposed multiplied by their average exposures, through application of the collective dose concept.

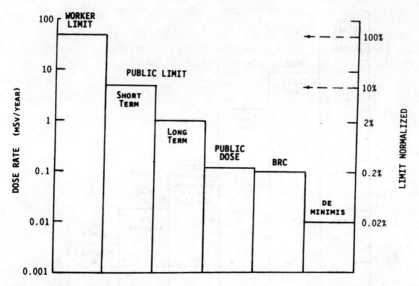

Fig. 4. Current Radiation Standards

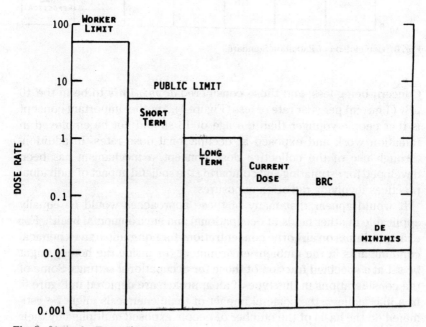

Fig. 5. Limits for Toxic Chemicals

Consideration might also be given to the application of an ALARA-like criterion in the control of such contaminants. In fact, now that these concepts have been so successfully applied in the development of standards for the control of radiation sources, perhaps the next step should be for radiation protection personnel to share these concepts with their fellow professionals who have the responsibility for health protection in other fields of occupational and environmental health.

References

1. Shakespeare, W., "As you like it," *Bartlett's Familiar Quotations*, (Pocket Books, Inc.) (1953).
2. Eisenbud, M., *Environment, Technology, and Health: Human Ecology in Historical Perspective* (New York University Press, New York) (1978).
3. Mutscheller, A., "Physical standards of protection against roentgen-ray dangers," American Journal of Roentgenology and Radium Therapy, Vol. 13, Issue 1, pages 65–70 (1925).
4. Upton, A.C., "The biological effects of low-level ionizing radiation," Scientific American, Vol. 246, No. 2 (1982).
5. Muller, J.J., "Radiation and human mutation," Scientific American, Vol. 193, No. 5 (1955).
6. NCRP. National Committee on Radiation Protection and Measurements. *Radium Protection for Amounts Up to 300 Milligrams*, Handbook 18, National Bureau of Standards (U.S. Government Printing Office, Washington) (1934).
7. NCRP. National Council on Radiation Protection and Measurements. *Safe Handling of Radioactive Luminous Compounds*, NCRP Report 5, originally published as Handbook 27, National Bureau of Standards (U.S. Government Printing Office, Washington) (1941).
8. NAS/NRC. National Academy of Sciences/National Research Council. *Summary Report by the Committee on the Biological Effects of Atomic Radiation (BEAR)* (National Academy Press, Washington) (1960).
9. NAS/NRC. National Academy of Sciences/National Research Council. *A Report to the Public by the Committee on Biological Effects of Atomic Radiation (BEAR)* (National Academy Press, Washington) (1956).
10. MRC. Medical Research Council. *The Hazards to Man of Nuclear and Allied Radiations*. Report No. Cmd. 9780 (Her Majesty's Stationary Office, London) (1956).
11. NCRP. National Council on Radiation Protection and Measurements. *Addendum to National Bureau of Standards Handbook 59, Permissible Dose from External Sources of Ionizing Radiation* (National Council on Radiation Protection and Measurements, Bethesda, Maryland) (1958).
12. NCRP. National Council on Radiation Protection and Measurements. *Maximum Permissible Body Burdens and Maximum Permissible Concen-*

trations of Radionuclides in Air and Water for Occupational Exposure, NCRP Report No. 22, originally published as Handbook 69, National Bureau of Standards (U.S. Government Printing Office, Washington) (1959).
13. NCRP. National Council on Radiation Protection and Measurements. *Permissible Dose From External Sources of Ionizing Radiation*, NCRP Report No. 17, originally published as Handbook 59, National Bureau of Standards (U.S. Government Printing Office, Washington) (1954).
14. ICRP. International Commission on Radiological Protection. *Recommendations of the International Commission on Radiological Protection*, ICRP Publication 1, Adopted September, 1958 (Pergamon Press, New York) (1959).
15. FRC. Federal Radiation Council. *Background Material for the Development of Radiation Protection Standards*, Report No. 1 (U.S. Environmental Protection Agency, Washington) (1960).
16. NRC. U.S. Nuclear Regulatory Commission. *Standards for Protection Against Radiation, Title 10, Part 20, Code of Federal Regulations* (U.S. Government Printing Office, Washington).
17. UNSCEAR. United Nations Scientific Committee on the Effects of Atomic Radiation. *Report of the United Nations Scientific Committee on the Effects of Atomic Radiation*, General Assembly Office Records: Seventeenth Session, Supplement No. 16 (A/5814) (United Nations, New York) (1961).
18. ICRP. International Commission on Radiological Protection. "The evaluation of risks from radiation," ICRP Publication 8, Health Phys. **12,** 239–302 (1966).
19. NCRP. National Council on Radiation Protection and Measurements. *Basic Radiation Protection Criteria*, NCRP Report No. 39 (National Council on Radiation Protection and Measurements, Bethesda, Maryland) (1971).
20. UNSCEAR. United Nations Scientific Committee on the Effects of Atomic Radiation. *Sources and Effects of Ionizing Radiation*, Report of the Scientific Committee on the Effects of Atomic Radiation, Official Records of the General Assembly, Thirty-Second Session, Supplement No. 40 (United Nations, New York) (1977).
21. NAS/NRC. National Academy of Sciences/National Research Council. *The Effects on Populations of Exposure to Low Levels of Ionizing Radiation*, Report of the Advisory Committee on the Biological Effects of Ionizing Radiation, BEIR III Report (National Academy Press, Washington) (1980).
22. Lewis, E.B., "Leukemia and ionizing radiation," Science **125,** 965–972 (1957).
23. NAS/NRC. National Academy of Sciences/National Research Council. *The Effects on Populations of Exposure to Low Levels of Ionizing Radiation*, Advisory Committee on the Biological Effects of Ionizing Radiation, BEIR I Report (National Academy Press, Washington) (1972).
24. UNSCEAR. United Nations Scientific Committee on the Effects of Atomic Radiation. *Ionizing Radiation: Levels and Effects*, Report of the Scientific Committee on the Effects of Atomic Radiation, Official Records of the

General Assembly, Twenty-Seventh Session, Supplement No. 25, Two Volumes (United Nations, New York) (1972).
25. ICRP. International Commission on Radiological Protection. *Recommendations of the International Commission on Radiological Protection*, ICRP Publication 26, Annals of the ICRP, Vol. 1, No. 3 (Pergamon Press, New York) (1977).
26. ICRP. International Commission on Radiological Protection. *Problems in Developing an Index of Harm*, ICRP Publication 27, Annals of the ICRP, Vol. 1, No. 4 (Pergamon Press, New York) (1977).
27. NCRP. National Council on Radiation Protection and Measurements. *Ionizing Radiation Exposure of the Population of the United States*, NCRP Report No. 93 (National Council on Radiation Protection and Measurements, Bethesda, Maryland) (1987).
28. NCRP. National Council on Radiation Protection and Measurements. *Recommendations on Limits for Exposure to Ionizing Radiation*, NCRP Report No. 91 (National Council on Radiation Protection and Measurements, Bethesda, Maryland) (1987).
29. IAEA. International Atomic Energy Agency. *Principles for the Exemption of Radiation Sources and Practices from Regulatory Control*, Safety Series No. 89 (International Atomic Energy Agency, Vienna) (1988).
30. EPA. U.S. Environmental Protection Agency. *Environmental Radiation Protection Standards for Nuclear Power Operations Title 40, Part 190, Code of Federal Regulations* (U.S. Government Printing Office, Washington).
31. EPA. U.S. Environmental Protection Agency. *Environmental Standards for the Management and Disposal of Spent Nuclear Fuel, High-Level and Transuranic Radioactive Wastes, Title 40, Part 191, Code of Federal Regulations* (U.S. Government Printing Office, Washington).
32. NRC. U.S. Nuclear Regulatory Commission. *Numerical Guides for Design Objectives and Limiting Conditions for Operation to Meet the Criteria As Low As Reasonably Achievable For Radioactive Material In Light-Water-Cooled Nuclear Power Reactor Effluents, Appendix I, Title 10, Part 50, Code of Federal Regulations* (U.S. Government Printing Office, Washington).
33. *Radiation Protection Guidance to Federal Agencies for Occupational Exposure*, Presidential Documents (U.S. Government Printing Office, Washington) (1987).
34. Mills, W.A., Flack, D.S., Arsenault, F.J. and Conti, E.F. *A Compendium of Major U.S. Radiation Protection Standards and Guides: Legal and Technical Facts* (Committee on Interagency Radiation Research and Policy Coordination, Washington) (1988).
35. *Radon Pollution Control Act of 1988*, Congressional Record, U.S. House of Representatives, pages H 9624–9637 (U.S. Government Printing Office, Washington) (1988).
36. Kocher, D. C., "Review of radiation protection and environmental radiation standards for the public," Nuclear Safety **29**, No. 4, 463–475 (1988).

Discussion

JOHN CAMERON: (University of Wisconsin): I've enjoyed your talk very much. I'd like to suggest, in an age of sophistication and maturity that we work better at communicating with the public, and develop a unit, such as radiation risk per year from natural radiation, which is something that should be understandable to them rather than millisievert's effective and so on. For at the moment the public gets their education from our television systems and I think you know that isn't very good.

DADE MOELLER: Thank you; I agree.

AL TSCHAECHE (WINCO): Dade, I think you have N minus one ages; there's one more.

DADE MOELLER: Thank you.

Al TSCHAECHE: And hopefully it will occur perhaps by the turn of the century; and that is the age of the acceptable risk. My question is how on earth are we going to get Congress to pass a number into law which will be an acceptable risk for all things not just radiation?

DADE MOELLER: You're correct; such a number would be very useful as a guide in all fields of occupational and environmental health.

JOE SOLDAT (PNL): We've tossed around in my establishment this thought of limiting collective dose depending on the number of people exposed, which is in essence putting a numerical value on the collective dose limit. The problem we see with that is that when you're exposing a small number of people you're allowing each individual to get a higher dose than we're exposing a large number of people to, which is philosophically okay but if I'm the guy that's getting an exposure in a small group I'm going to complain and say - "How come I'm allowed ten times the dose that the fellow over there is? I'm just as important or more so than he is. And I don't think it's fair."

DADE MOELLER: I think that's something we have to address and try to work out a system for handling.

JOE SOLDAT: Yes.

PAUL STANSBURY (GE Nuclear Energy): On March 7th of this year the Environmental Protection Agency proposed some radionuclide emission standards under the Clean Air Act. They proposed dose limits between 10 millirem and .03 millirem per year. Would you care to comment on those numbers as they relate to the NCRP's concepts of BRC and *de minimis*, please?

DADE MOELLER: About the only comment I could make is that we should place those proposed limits on the chart that I showed, and see where they fall, comparing radon and high level waste repositories and so forth and perhaps by doing that we can put the numbers in perspective and help EPA to understand where they fit.

EUGENE SAENGER (University of Cincinnati): Dade, recently there's been a lot of comment about various environmental agents, toxins, whatever, and there's a question of whether we can afford this. As we keep running these numbers down, which is the natural direction for all regulators, one begins to wonder how do we determine that we're not spending more than is really necessary in this particular effort, could you comment on that?

DADE MOELLER: I don't really know what response to give. I would suggest that the NCRP set up a scientific committee to handle this.

MERRIL EISENBUD: I'd like to take a whack at it. In fact, some of you may have seen a letter I sent to Science on this a few months ago. If you take Bernie Cohen's data that you can prevent death from cervical cancer by spending $25,000 for education and more available screening, then you can save a death from smoke inhalation by spending on an average $40,000 worth of smoke detectors. You find that as you get into the more sophisticated technology such as chemical technologies that the cost per death averted runs up to about a hundred to two hundred million. But when you get into radiation and take as the boundary case the recent problems the TMI people have been having with the discharge of tritiated water into the Susqehanna River, they're spending ten million dollars to avert a dose equivalent of something equivalent to four minutes of natural radiation exposure and that works out to seventeen trillion dollars per death averted.

BOB VANDERSLICE (EPA): How do you reconcile your approach for regulation with the fact that the different agencies have different regulatory mandates given them from Congress?

DADE MOELLER: I think that this relates back to an earlier comment that if we could have Congress give us an acceptable risk number that all agencies would apply uniformly; this would be very helpful, I agree.

Scientific Session

Scientific Basis

Gilbert Beebe
Chairman

Status of Somatic Risk Estimation

William J. Schull
University of Texas
Heath Science Center
Houston, TX

Hiroo Kato
Yukiko Shimizu
Yasuhiko Yoshimoto
Masanori Otake
Radiation Effects Research Foundation
Hiroshima, Japan

Introduction

Some ten years ago, as most of you are aware, there began a reassessment of the exposures of the survivors of the atomic bombing of Hiroshima and Nagasaki prompted by certain physical and biological incongruities in the data associated with the T65 doses. This reevaluation culminated in a new system of dosimetry, the so-called DS86 (1). As originally installed, this system afforded a means to compute doses for those survivors with the requisite shielding information by directly modeling the circumstances attending their exposure, and to estimate doses indirectly for a substantial number of survivors without detailed shielding data. Collectively, doses were estimated for some 82% of the exposed members of the Life Span Study sample, but a lesser fraction in other study groups, such as the parents of the F_1.

Recently, the system was extended to include direct computation of doses for those survivors in Nagasaki who were terrain shielded, or exposed in factories. This involves approximately 1000 individuals in

TABLE 1—*Current estimates of the risk of site-specific malignancies based upon mortality in the Life Span Study sample in the years 1950 to 1985. These estimates are based upon shielded kerma, a linear response model over the whole dose range, and only those members of the sample on whom DS86 doses exist (both cities, sexes, and all ages ATB combined). This table has been adapted from Table 6 in Shimizu et al. (5)*

Site of cancer	Excess relative risk per gray	Excess deaths per 10^4 PYGy	Attributable risk (%)
Leukemia	3.97 (2.89, 5.39)	2.30 (1.88, 2.73)	56.6 (46.3, 67.1)
All except leukemia	0.30 (0.23, 0.37)	7.49 (5.90, 9.15)	8.0 (6.3, 9.8)
Esophagus	0.43 (0.09, 0.92)	0.34 (0.08, 0.67)	12.8 (3.0, 25.0)
Stomach	0.23 (0.13, 0.34)	2.09 (1.20, 3.06)	6.4 (3.6, 9.3)
Colon	0.56 (0.25, 0.99)	0.56 (0.26, 0.91)	15.2 (7.0, 24.9)
Lung	0.46 (0.25, 0.72)	1.26 (0.70, 1.89)	11.6 (6.5, 17.4)
Breast	1.02 (0.48, 1.76)	1.04 (0.53, 1.61)	22.4 (11.5, 35.0)
Ovary	0.80 (0.14, 1.85)	0.45 (0.09, 0.89)	18.6 (3.6, 37.1)
Bladder	1.06 (0.46, 1.09)	0.56 (0.27, 0.90)	23.4 (11.2, 37.4)
Multiple myeloma	1.89 (0.56, 4.45)	0.22 (0.08, 0.39)	32.9 (11.5, 59.8)

( ): 90% confidence interval

the Life Span Study sample. Rules have also been formulated for the indirect estimation of doses for a larger number of survivors exposed beyond 2000 m, largely individuals in the open or with no shielding information. While this has further reduced the number of survivors without DS86 doses, there remain at present classes of individuals on whom doses are still not available.

These dosimetric developments coupled with further follow-up have led to a reevaluation of the effects of exposure on the survivors, and out of this effort has grown a series of new estimates of cancer, risk. It is these estimates, which serve as the bases for the risk projections that have recently been published by UNSCEAR (2), the United Kingdom's National Radiological Protection Board (3) as well as those of the National Academy of Science's Committee on the Biological Effects of Ionizing Radiation (BEIR-V) (4) to appear shortly.

Survivors Exposed Postnatally

Cancer has been, and continues to be the most conspicuous finding among survivors exposed postnatally to the atomic bombing of Hiroshima and Nagasaki. The risk coefficients associated with the new doses are summarized site-specifically in Tables 1 and 2 for shielded kerma and organ absorbed dose, respectively. Since these values have been previously published and discussed on several previous occasions (5) (6), we will not dwell on them here; however, certain points made earlier

TABLE 2—*Current estimates of the risk of site-specific malignancies based upon mortality in the Life Span Study sample in the years 1950 to 1985. These estimates are based upon organ doses, a linear response model over the whole dose range, and only those members of the sample on whom DS86 doses exist (both cities, sexes, and all ages ATB combined). This table has been adapted from Table 7 in Shimizu et al. (5)*

Site of cancer	Excess relative risk per gray	Excess deaths per 10^4 PYGy	Attributable risk (%)
Leukemia	5.21 (3.83, 7.14)	2.94 (2.43, 3.49)	58.6 (48.4, 69.5)
All except leukemia	0.41 (0.32, 0.51)	10.13 (7.96, 12.44)	8.1 (6.4, 10.0)
Esophagus	0.58 (0.13, 1.24)	0.45 (0.10, 0.88)	13.0 (3.0, 25.5)
Stomach	0.27 (0.14, 0.43)	2.42 (1.26, 3.72)	5.7 (3.0, 8.7)
Colon	0.85 (0.39, 1.45)	0.81 (0.40, 1.30)	16.3 (8.0, 26.2)
Lung	0.63 (0.35, 0.97)	1.68 (0.97, 2.49)	12.3 (7.2, 18.3)
Breast	1.19 (0.56, 2.09)	1.20 (0.61, 1.91)	22.1 (11.3, 35.0)
Ovary	1.33 (0.37, 2.86)	0.71 (0.22, 1.32)	22.3 (6.9, 41.4)
Bladder	1.27 (0.53, 2.37)	0.66 (0.31, 1.12)	21.5 (9.8, 35.7)
Multiple myeloma	2.29 (0.67, 5.31)	0.26 (0.09, 0.47)	31.8 (11.0, 57.6)

( ): 90% confidence interval

warrant reiteration. First, it should be noted that the risks when based on organ absorbed dose and expressed as excess deaths per 10^4 PYGy are generally altered by no more than 20% as contrasted with the comparable T65 estimate (the breast is one conspicuous exception); whereas the estimates based on shielded kerma change more, 40–60%. These values reflect the fact that the DS86 kerma doses are lower than the T65, on average, but the organ transmission factors are generally higher.

Second, although a linear dose-response model appears to be an acceptable and simple descriptor of the occurrence of most non-leukemic malignancies, other models, particularly a linear-quadratic, will often fit the data almost as well, if not as well. Further years of follow-up may clarify this matter, but perhaps one should not be overly sanguine. Unfortunately, the data are, and are likely to remain least informative where the regulatory concern is the greatest, that is, at doses of less than 0.10 Gy (7). Perforce, models are critical here, and the appropriate one will almost certainly remain contentious, see Gofman (8), and for a rebuttal, Muirhead and Butland (9).

Third, there now seems little likelihood of being able to estimate the relative biological effectiveness of neutrons based on the experiences of the survivors at least insofar as cancer is concerned. Values of 1 to 20 or so are consistent with the observed data.

Finally, the differences that have obtained in past analyses between the two cities are no longer significant statistically although for the same dose the risk remains higher in Hiroshima than Nagasaki. When coupled with other findings with regard to epilation (10), chromosomal abnormalities (11) and presumed somatic mutations, for example, this suggests a continuing need to search for alternate explanations for the city differences heretofore ascribed to the presumed differences in neutron flux. It has been suggested in this regard that possibly the yield of the Hiroshima weapon was actually somewhat greater than the consensus value of 15 kilotons.

It warrants noting that the introduction of the new dosimetry impinges on virtually all previous studies of the Foundation. All of the incidence data, such as that on leukemia or breast cancer, must be examined anew, as must the various case-control studies. Obviously, there remains an enormous task to reanalyze these in the light of the revised doses. While it is not anticipated that these reanalyses will alter the findings dramatically, clearly several more years may elapse before we know fully the implications of the new dosimetry.

Projections of risk

A variety of agencies and institutions, national and international, are presently engaged in the projection of lifetime risks of cancer based on these new estimates, but as yet only two have published their deliberations, namely, the United Nation's Scientific Committee on the Effects of Atomic Radiation (2), and the United Kingdom's National Radiological Protection Board (3). These assessments differ in their intent and methods although they share a number of particulars, not the least of which are the Japanese risk estimates themselves. However, the UNSCEAR projections have been deliberately made simple, for it was presumed that more complex models or more effort to adjust for the known shortcomings in the data would have made the results progressively more particular and less and less applicable to the broad community of nations that the Committee serves. No attempt was made to correct for the underreporting of cancer deaths on death certificates, since this could vary greatly among nations, nor were the age-specific risk coefficients "smoothed", nor were the relative risks presumed to decline with time.

Table 3 contrasts the lifetime projection of fatal cancers among the Japanese made by UNSCEAR with the projection for the United Kingdom made by the NRPB. Both sets of projections assume a linear dose-response relationship for cancers other than leukemia, and are based on the relative (or multiplicative) risk projection model. The latter seems more consistent with the experience of the survivors where the absolute excess risk of all cancers except leukemia continues to increase over time whereas the relative risk appears to be constant or nearly so, save

TABLE 3—*Estimated lifetime excess cancer risks in a population of all ages and both sexes associated with exposure to low LET radiation (adapted from UNSCEAR (2) Tables 62 and 69, and NRPB (3) Table 3).*

Cancer type	Fatal cancer risks, Japan	10^{-2} Gy^{-1} (low LET) United Kingdom
Leukemia	1.00	0.84
Other malignancies	9.70	11.38
Total	10.70*	12.22*
	Some specific sites	
Breast	0.30**	1.1**
Lung	1.51	3.5
Colon	0.79	1.1
Stomach	1.26	0.73

* These values exclude cancers of the thyroid, bone and liver for which UNSCEAR did not make specific projections.
**These UNSCEAR projections are based on an age-constant risk coefficient whereas the NRPB used age-specific risk coefficients.

possibly for the youngest age ATB group, that is, those exposed in the first decade of life. Although the details of the projections differ somewhat, as previously said, both are for a general population, and use age-specific risk coefficients. Note that the specific rates given in Table 3 in the instance of the UNSCEAR projections are based on an age-constant risk coefficient. It is important to bear in mind too that the background rates are different, in one instance those for the contemporary population of Japan (1982) and in the other the United Kingdom (1985). There is a general correspondence between these projections, see also those of Preston and Pierce (12) and Charles and Little (13), despite the differences in methods, that is heartening, but could be merely fortuitous.

It should be noted that the differences between these projections and earlier ones (e.g., UNSCEAR 1977 (14) lifetime estimate of about 2.5 10^2 Gy^{-1}) stem less from the changes in dose than from the use of the multiplicative rather than the additive projection model and the further refinement in the risk coefficients attributable to a longer period of surveillance.

The values cited in Table 3 are assumed to be strictly applicable only to doses of about 0.5 Gy or more received at a high dose rate. At lesser doses, and low dose rates the risk is assumed to be less. Although UNSCEAR favored no single dose rate effectiveness factor (DREF), it suggests that the value lies between 2 and 10, probably closer to the former than the latter. The National Radiological Protection Board in its estimates have used values of two for the breast, and three for other cancers.

New complexities in the projection of risk

Previous analyses of non-cancer mortality among the survivors have failed to disclose systematic effects of exposure on causes of death other than cancer, see, *e.g.*, Kato, Brown, Hoel and Schull (15). However, this no longer appears to be the case. Shimizu and her colleagues (16) now find that mortality from all diseases except neoplasms in the years 1950–85 exhibits a linear-quadratic dose-response with excess risks apparent at doses of 2 or 3 Gy and over, as contrasted with the 0 dose group. This increase is most readily demonstrable in the recent time periods and in the younger age ATB groups (less than 40) suggesting a sensitivity for these groups. When more specific, smaller disease categories are studied, the increasing trend with dose is statistically significant in only a few instances, but an excess in relative risk at doses of 2 Gy or more is seen in the majority. For most, but not all of these disease categories, a linear threshold model fits the data as well as any other. The estimated threshold is generally about 1.5 Gy.

These authors cautiously note that the data are still limited, the dose-response relationships far from clear, and that only further surveillance will provide firmer insights. However, if true in any of their particulars, these findings have important implications for the projection of cancer risks. Past projections have invariably assumed that deaths from competing causes of mortality are independent of exposure. If this is not true, projections will necessarily be more complex and tenuous than we have believed. Indeed, it may be possible to estimate the net cause-specific hazards, see, *e.g.*, Cornfield, (17) Tsiatis (18), and Schatzkin and Slud (19).

Survivors Exposed Prenatally

Developments over the past several years have disclosed risks associated with prenatal exposure to ionizing radiation more pervasive than previously thought. First, Yoshimoto and his colleagues (20) have shown an increased risk of malignant solid tumors among these survivors despite the fact that no significant increase was observed in childhood leukemia. Second, the continuing reassessment of brain damage among the prenatally exposed has not only sharpened our knoweldge of the period of maximum vulnerability, but has revealed evidence of more subtle forms of impairment.

TABLE 4—*Summary measures of cancer incidence (1950–84) in relation to prenatal exposure to A-bomb radiation based on the estimated DS86 dose absorbed by the mother's uterus and a linear dose-response model [Modified from Table V of (20)].*

	Relative risk (at 1 Gy)	Average excess risk (per 10^4 PYGy)	Attributable risk (%)
Follow-up 1950–84 all ages*			
Estimated value	3.77	6.57	40.9
95% lower bound	1.14	0.47	2.9
95% upper bound	13.48	14.49	90.2
Follow-up over age range 15–39 years**			
Estimated value	2.44	4.80	26.0
95% lower bound	0.71	−1.26	−6.8
95% upper bound	9.49	14.98	81.2

* Significant ($p = 0.03$)
**Not significant ($p = 0.16$)

Cancer

Over the period from 1950 through 1984, based on the absorbed dose to the mother's uterus as estimated by the DS86 doses, the relative risk of cancer at 1 Gy is 3.77 with a 95% confidence interval of 1.14–13.48. Among those survivors receiving 10 mGy or more, the average excess risk per 10^4 person-year-gray is 6.57 (0.07–14.49) and the attributable risk is 41% (2.9–90%). Moreover, not only did these cancers occur earlier in the 0.30+ Gy dose group than in the 0 dose group, but the incidence continues to increase, and the crude cumulative incidence rate, 40 years after the bombing, is 3.9-fold greater in the 0.30+ Gy group. These findings are summarized in Table 4, adapted from Yoshimoto, Kato and Schull (20).

The specific tumors that have been seen are, in the main (10 of 16 non-leukemic malignancies), the same ones known to be increased in frequency among the postnatally exposed survivors, i.e., bladder, breast, colon, ovary, stomach and thyroid. While the data are limited, they are consistent with a risk of cancers other than leukemia at least as high, and possibly higher than that seen in individuals exposed in the first decade of life whose own risk has heretofore been the highest of any age group of survivors. We do not yet know why the risk of leukemia in the prenatally exposed survivors is not commensurate with that seen in survivors 0–9 ATB.

Brain Damage

Quantitative risk estimates for radiation-related damage to the brain after prenatal exposure are of practical importance not because the

TABLE 5—*The effect on the developing brain of exposure to ionizing radiation in weeks 8–15 following fertilization*

Effect	Risk at 1 Gy	Comments
Severe mental retardation	Increased 50-fold	Risk rises from 0.8% at 0 Gy to 44% at 1 Gy
Intelligence test score	Decreased 24–33 pts*	This is a decline of about two standard deviations
School performance	Decreased 1.0–1.3 pts**	This is a fall from the class 50 percentile to the lowest 10 percentile
Seizures, unprovoked	Increased 20-fold	Risk rises from 0.9% at 0 Gy to 20% at 1 Gy

* Note these values do not represent the upper and lower confidence limits, but the range of central estimates based on samples including and excluding individuals known to be mentally retarded.

**Again this is not the confidence interval, but the range of central estimates seen over the four grades in school that have been studied.

population involved is necessarily large but rather because the probability of a deleterious effect appears so high. However, the human data on which to base such estimates, provocative though they are, are limited and imperfect. Four types of observations are available, namely, (1) the frequency of clinically recognized severe mental retardation, (2) the diminution of intelligence as measured by conventional intelligence tests, (3) scholastic achievement in school, and finally, (4) the occurrence of unprovoked seizures. All are interrelated to some degree, and each has its own short-comings. Although cognizant of these and other difficulties inherent in the interpretation of the available information, until such time as more direct measures of brain damage, such as cell death or impaired cell migration, are available, these observations are the only ones on which risk estimates, can be based. Anecdotal clinical evidence is of little assistance and experimental data, though important qualitatively, provide an uncertain basis for quantitative estimates of prenatal risks in the human.

Reevaluation of the Japanese atomic bomb survivor data has provided a new perspective on the periods of sensitivity of the developing brain to radiation-related damage, the possible nature of the dose-response relationship, and some insight into the biologic bases of the events that are seen. These findings, as they specifically concern risk estimation, are as follows (see Table 5):

The period of maximum vulnerability to radiation appears to be the time from approximately the beginning of the eighth through the fifteenth week after fertilization, that is, within the interval when the greatest proliferation of neurons and their migration to the cerebral cortex occur. A period of lesser vulnerability occurs in the succeeding period from the sixteenth through the twenty-fifth week after fertilization. The later period accounts for about a fourth of the apparently radiation-related cases of severe mental retardation. The least vulnerable time is apparently the initial eight weeks after fertilization; no radiation-related cases of severe mental retardation have been seen in this developmental stage. This should not be construed as evidence that brain damage does not occur at this time, for it may, but be incompatible with continued survival to ages at which mental retardation can be recognized.

Within the period of maximum vulnerability, the simplest statistical model consistent with the data appears to be a linear one without threshold. The slope of this relationship, based on the supposition that the occurrence of mental retardation is binomially distributed, corresponds to an increase in frequency of severe mental retardation of 0.44 per Gy (95% CI: 0.26–0.62). Thus, the frequency of severe mental retardation rises from about one case per hundred individuals exposed to less than 0.01 Gy to approximately 44 cases per hundred at an exposure of 1 Gy. Uncertainties in assigning some exposures to a defined stage of pregnancy or the exclusion of those cases of mental retardation with probable non-radiation related etiology have been found to have little effect on this risk estimate.

The data on intelligence tests and school performance suggest the same two gestational periods of vulnerability to radiation, the first period showing the greatest sensitivity. More importantly, these data suggest a continuum of effects on the developing brain of exposure to ionizing radiation; indeed, the downwards shift we see in the distribution of IQ scores with increasing exposure predicts reasonably well the actual increase in severe mental retardation that has been observed. This suggests, in turn, that the impact of exposure to ionizing radiation will be related to where in the normal continuum of cortical function an individual would have resided if unexposed. Simply put, the loss, say, of 5 IQ points in one destined to have an IQ of 140 would hardly be handicapping, but a similar loss at an IQ of 75 could result in mental retardation.

Seizures are a frequent sequela of impaired brain development, and therefore, could be expected to affect more children with radiation-related brain damage than children without. Dunn and her colleagues (21) have described the incident, and type, of seizures among survivors prenatally exposed to the atomic bombing of Hiroshima and Nagasaki, and their association with specific stages of pre-natal development at the time of irradiation. Histories of seizures were obtained at biennial routine clinical examinations starting at the age of two years. These clinical records were used to classify seizures as febrile or unprovoked (without precipitating cause).

Seizures were not recorded among individuals exposed 0–7 weeks after fertilization at doses higher than 0.10 Gy. After irradiation at 8–15 weeks after fertilization, the incidence of seizures was highest among individuals with doses exceeding 0.10 Gy and was linearly related to the level of fetal exposure. This obtains for all seizures without regard to the presence of fever or precipitating causes, and for unprovoked seizures. When the 22 cases of severe mental retardation were excluded, the increase in seizures was only suggestively significant and then only for unprovoked seizures. After exposure at later stages of development, there was no increase in recorded seizures.

The risk ratios for unprovoked seizures, following exposure within the 8th through the 15th week after fertilization, are 4.4 (90% confidence interval: 0.5–40.9) after 0.10–0.49 Gy and 24.9 (4.1–191.6) after 0.50 or more Gy when the mentally retarded are included, and 4.4 (0.5–40.9) and 14.5 (0.4–199.6), respectively, when they are excluded.

It is not clear which of these analyses, that based on the inclusion or the exclusion of the mentally retarded, should be given the greater weight. The answer hinges ultimately on the mechanisms underlying the occurrence of seizures and mental retardation following prenatal exposure to ionizing radiation, and these are presently unknown. If seizures can arise by two independent mechanisms, both possibly dose related, one of which causes seizures and the other causes mental retardation in some individuals who are then predisposed to develop seizures, the mentally retarded must necessarily be excluded if one is to explore the dose-response relationship associated with the first mechanism. If, however, mental retardation and seizures arise from a common brain defect, which manifests itself in some instances as mental retardation and in others as seizures, then the mentally retarded should not be excluded. At present the only evidence arguing for a common radiation-related developmental defect is the occurrence of ectopic gray areas in some instances of both disorders. But, even this evidence, is difficult to put into perspective, for while it is know that ectopic gray areas occur among some of the radiation-related instances of mental retardation, the observation of ectopia in individuals with seizures is based on other studies. There has been no investigation of the frequency of occurrence of ectopic gray areas among the prenatally exposed with seizure but no mental retardation.

At present, there is no evidence of radiation-related cerebellar damage without concomitant damage to the cerebrum in the survivors of the atomic bombing of Hiroshima and Nagasaki exposed prenatally. It may be difficult to identify such damage for several reasons. First, Purkinje cells, the only efferent neurons in the cerebellum, are proliferating and migrating in the same developmental period as the neuronal cells that populate the cerebral cortex, and thus, damage to precursors or differentiated Purkinje cells would occur at the same time and may be inseparable from damage to those cells that give rise to the cerebral cortex. Second, the granular neurons, the most numerous nerve cells in the cerebellum, retain their proliferative abilities after birth and could, in theory, repopulate areas of the developing cerebellum damaged by radiation. To the extent that this occurs, granular cell damage might be mitigated. Estimates of the risk of damage to the cerebellum following prenatal exposure, based on fixed or progressive neurological deficit, are presently not possible.

Uncertainties

Many uncertainties are associated with these estimates of risk. They include the limited nature of the data, especially on mental retardation and convulsions, the appropriateness of the comparison group, errors in the estimation of the tissue absorbed doses and the prenatal ages at exposure, and other confounding factors in the post-bomb period, including nutrition and disease, which could play a role. Time precludes a measured consideration of each of these; suffice it to state that no fully satisfactory assessment of their contribution can be made at this late date. Given the present uncertainties, since most of these extraneous sources of variation would have a greater impact at high than low doses, and thus, produce a concave upwards dose-response function, the prudent course would be to assume that the dose-response relationship is not materially altered other than additively by these potential confounders. This would have the effect of overestimating the risk at low doses where greatest regulatory concern exists.

Three issues do warrant consideration here; these include the shape of the dose-response function, the existence of a threshold in the dose-response, and the effects of dose fractionation.

The dose-response function: — Within the period of maximum vulnerabilty, virtually without exception, the data previously presented can be satisfactorily approximated by more than one dose-response function, generally a linear or a linear-quadratic model. Given that neuronal death, mismanaged migration, and faulty synaptogenesis could all play a role in the occurrence of mental retardation or cortical dysfunction more generally, and that each could have its own different dose-response relationship, there is little or no prior basis for presuming that one or the other of these models better describes the fundamental biological events involved. The "true" model, therefore, remains a matter of conjecture, and it seems unlikely that epidemiological studies alone will ever be able to determine what the "true" model may be. Of necessity the estimation of risk must rest on a series of considerations, not all of which are biological. Most importantly, the risk estimated should be a prudent one, minimizing risk wherever such exists. This argues for the use of a linear dose-response since presumably at lower doses, where the evidence of an effect is weakest, risk is apt to be overestimated.

Is there a threshold? — Although a linear or a linear-quadratic dose-response relationship describes the observed frequency of severe mental retardation in the 8th through the 15th week adequately, there could be a threshold with the DS86 dosimetry. As Otake, Yoshimaru and Schull (22) have shown, the estimation of the value of this presumed threshold is not straightforward. When all of the cases of mental retardation are included in the analysis, the lower bound of the estimate of the threshold includes zero, that is to say, a threshold cannot be shown to exist by statistical means. But if the two cases of Down's syndrome in the 8–15 week period are excluded, the 95% lower bound of the threshold appears to range from 0.12 to 0.23 Gy. It should be noted, however, that the imposition on the data of a linear model with a threshold gives rise to a rate of increase with dose that predicts virtually every fetus exposed to one gray or more will be retarded. This is at variance with the actual observations, but this would not necessarily be true of a curvilinear model with a threshold. The DS86 dosimetry suggests a threshold in the 16–25 week period of 0.21–0.70 Gy.

Dose fractionation: — Little is known about the effects on the developing human embryo and fetus of chronic or fractionated exposures to ionizing radiation. Given the complexity of brain development and the differing durations of specific developmental phenomena, it is reasonable, however, to assume that dose fractionation will have some effect. The hippocampus, for example, and the cerebellum continue to have limited neuronal multiplication, and migration does occur in both organs. Changes continue in the hippocampus and cerebellum into the first and second years of life. Continuing events such as these may show dose-rate effects differing from those associated with the multiplication of the cells of the ventricular and subventricular areas of the cerebrum, or the migration of neurons to the cerebral cortex.

Most of the information available on the effects of dose rate involves the experimental exposure of rodents, and must be interpreted with due regard to the differences between species in developmental timing and rates relative to birth. Brizzee and Brannon, (23) for example, see also Jacobs and Brizzee, (24) have examined cell recovery in the fetal brain of rats. The incidence and severity of tissue alterations generally varied directly with dose, and were clearly greater in single dose than in split dose groups with the same total exposure. These authors observed that "The presence of a threshold (shoulder) zone on the dose-response curve in the split-dose animals suggests that cell recovery occurred in some degree in the interval between the two exposures."

To my view, at least, the evidence which has emerged from the studies of brain damage among the prenatally exposed survivors present a biological coherence that would have been difficult to anticipate. Admittedly, the shape of the dose-response and the existence of a threshold are unresolved. Neither of these concerns are likely to be clarified by additional epidemiological studies, nor selective consideration of portions of the data now available, nor further manipulation of the doses. Answers can only come through a better understanding of the basic molecular and cellular processes that are involved.

As yet, unfortunately, we know far too little about these processes to do more than speculate on the origin of the effects that are seen; however, the brains of six of the 30 mentally retarded individuals in the clinical sample, as well as a seventh retarded individual not in the sample, have been examined either at autopsy or through magnetic resonance imaging. The findings on these individuals are informative and suggest that errors in migration are common. They are seen in no less than five of the six cases exposed in the 8 to 15th week window. For example, coronal sections of the cerebrum of one of the mentally retarded individuals, a male, with a brain weighing 840 grams, who died at age 16 of acute meningitis (25) and came to autopsy, revealed massive ectopic gray matter around the lateral ventricles. Histologically there was an abortive laminar arrangement of nerve cells within the heterotopic gray areas, imitating the normal laminar arrangement of the cortical neurons. The cerebellum and hippocampi were normal histologically. Similarly, in four of five individuals on whom magnetic resonance imaging has occurred there is either direct or indirect evidence of mismanaged neuronal migration. In two cases exposed at the eighth or ninth week following fertilization, large areas of periventricular ectopic gray matter, comparable to those in the autopsied case, are seen. Two individuals exposed in the 12th to 13th week who do not show readily recognizable ectopic gray areas, do show mild macrogyria which implies some impairment of the migration of neurons to the cortical zone. Both of these individuals have cerebellar anomalies. Why the immature neuronal cells failed to migrate is not clear, but it could reflect changes in the intracellular adhesiveness which plays such a large role in their ultimate cortical positioning.

Although it has been common to view brain damage, and in particular mental retardation as a non-stochastic phenomenon, and therefore to suppose a threshold exists, there is no compelling molecular or cellular argument to support this. Chance or stochastic events can play a large role in morphogenesis as Kurnit and his colleagues (26) have shown; they have demonstrated that even when migration and cell division times are constant, variation in intracellular adhesiveness can alter the liability to abnormality. Their conclusions rest, however, on elaborate simulations, and it is to be hoped that future experimental studies will attempt to address more directly some of the issues the epidemiological data have raised. For example, *in vitro* studies of early brain development in explants or systems such as that used by Fushiki (27) could provide information, preliminary at least, on the effects of doses of a few rad on neuronal migration. Similarly, given the number of surface and subsurface proteins now recognized to be involved in the migratory process, it should be possible to determine whether specific ones are altered, permanently or even transitorily, by exposure to ionizing radiation. Clearly, there is much still to be learned, but recent advances in the neurosciences suggest a better understanding will be forthcoming.

References

1. RERF. Radiation Effects Research Foundadion. *U.S. Japan Joint Reassessment of Atomic Bomb Radiation Dosimetry in Hiroshima and Nagasaki: Final Report,* Roesch, W. C., Ed. (Radiation Effects Research Foundation, Hiroshima) (1987).
2. UNSCEAR. United Nations Scientific Committee on the Effects of Atomic Radiation. *Sources, Effects and Risks of Ionizing Radiation,* Report to the General Assembly, with annexes (United Nations, New York) (1988).
3. Stather, J. W., Muirhead, C. R., Edwards, A. A., Harrison, J. D., Lloyd, D. C. and Wood, N. R. *Health Effects Models Developed from the 1988 UNSCEAR Report* Document NRPB-R226 (National Radiological Protection Board, Chilton, Didcot, Oxfordshire, United Kingdom).
4. NAS/NRC. National Academy of Sciences/National Research Council. BEIR V (National Academy Press, Washington) (in press).
5. Shimizu, Y., Kato, H., Schull, W. J., Preston, D. L., Fujita, S. and Pierce, D. A. *Life Span Study Report 11. Part 1. Comparison of Risk Coefficients for Site-Specific Cancer Mortality Based on the DS86 and T65D Shielded Kerma and Organ Doses,* Technical Report 12-87 (Radiation Effects Research Foundation, Hiroshima) (1987).

6. Schull, W. J., Kato, H. and Shimizu, Y. "Implications of the New Dosimetry for Risk Estimates," pages 218–226 in *New Dosimetry at Hiroshima and Nagasaki and Its Implications for Risk Estimates*, Proceedings of the Twenty-third Annual Meeting of the National Council on Radiation Protection and Measurements (National Council on Radiation Protection and Measurements, Bethesda, Maryland) (1988).
7. Land, C. E. "Estimating cancer risks from low doses of ionizing radiation," Science **209**, 1197–1203 (1980).
8. Gofman, J. W. "Warning from the A-bomb study about low and slow radiation exposures," Letter to the Editor, Health Phys. **56**, 117–118 (1989).
9. Muirhead, C. R. and Butland, B. K. "Dose-response analyses for the Japanese A-bomb survivors," Letter to the Editor, Health Phys. (in press).
10. Stram, D. O. and Mizuno, S. "Analysis of the DS86 atomic bomb radiation dosimetry methods using data on severe epilation," Radiat. Res. **117**, 93–113, (1989).
11. Awa, A. "Chromosome aberration data for A-bomb dosimetry," page 185–203 in *New Dosimetry at Hiroshima and Nagasaki and its Implications for Risk Estimates*, Proceedings of the Twenty-third Annual Meeting of the National Council on Radiation Protection and Measurements (National Council on Radiation Protection and Measurements, Bethesda, Maryland) (1988).
12. Preston, D. L. and Pierce, D. A. "The effects of changes in dosimetry on cancer mortality risk estimates in the atomic bomb survivors," Radiat. Res. **114**, 437–466 (1987).
13. Charles, M. W. and Little, M. P. *Review of the Current Status of Radiation Risk Estimates*, Central Electricity Generating Board Research Report RD/B/6112/R89 (Central Electricity Generating Board, London) (1988).
14. UNSCEAR. United Nations Scientific Committee on the Effects of Atomic Radiation. *Sources and Effects of Ionizing Radiation*, 1977 Report to the General Assembly, with annexes (United Nations, New York) (1977).
15. Kato, H., Brown, C. C., Hoel, D. G. and Schull, W. J. "Studies of the mortality of A-bomb survivors, Report 7. Mortality 1950–78 Part 2. Mortality from causes other than cancer and mortality in early entrants," Radiat. Res. **91**, 243–64 (see also Technical Report 5-81 Radiation Effects Research Foundation, Hiroshima) (1982).
16. Shimizu Y., Kato, H., Schull, W. J. and Hoel, D. G. *Life Span Study Report 11. Part 3. Non-cancer mortality in the years 1950–85 based on the recently revised doses (DS86)*, Technical Report X-89 (Radiation Effects Research Foundation, Hiroshima) (1989).
17. Cornfield, J. "The estimation of the probability of developing a disease in the presence of competing risks," Amer. J. Public Health **47**, 601–607 (1957).
18. Tsiatis, A. "A nonidentifiability aspect of the problem of competing risks," Proc. Natl. Acad. Sci. USA **72**, 20–22 (1975).
19. Schatzkin, A. and Slud, E. "Competing risks bias arising from an omitted risk factor." Amer. J. Epidemiol. **129**, 850–856 (1989).

20. Yoshimoto, Y., Kato, H. and Schull, W. J. "Risk of cancer among children exposed in utero to A-bomb radiations 1950–84," Lancet ii 665–669 (see also Technical Report 4-88 Radiation Effects Research Foundation, Hiroshima) (1988).
21. Dunn, K., Yoshimaru, H., Otake, M., Annegers, J. F. and Schull, W. J. *Prenatal exposure to ionizing radiation and subsequent development of seizures*, Technical Report 5-88 (Radiation Effects Research Foundation, Hiroshima) (1988).
22. Otake, M., Yoshimaru, H. and Schull, W. J. *Severe Mental Retardation Among the Prenatally Exposed Survivors of the Atomic Bombing of Hiroshima and Nagasaki: A Comparison of the T65DR and DS86 Dosimetry Systems*, Technical Report 16-87 (Radiation Effects Research Foundation, Hiroshima) (1987).
23. Brizzee, K. R. and Brannon, R. B. "Cell recovery in foetal brain after ionizing radiation," Int. J. Radiat. Biol. **21**, 375–388 (1972).
24. Jacobs, L. A. and Brizzee, K. R. "Effects of total-body X-irradiation in single and fractionated doses on developing cerebral cortex in rat foetus," Nature **210**, 31–33 (1966).
25. Yokota, S., Tagawa, D., Otsuru, S., Nakagawa, K., Neriishi, S., Tamaki, H. and Hirose, K. "Tainai hibakusha ni mirareta shotosho no ichi bokenrei," Nagasaki Medical J. **38**, 92–95 (1963).
26. Kurnit, D. M., Layton, W. M. and Matthysse, S. "Genetics, chance and morphogenesis," Amer. J. Hum. Genet. **41**, 979–995 (1987).
27. Fushiki, S., Kawase, M., Fukuyama, R. and Fujita, S. "Morphogenesis *in vitro* of neural tube-like structures by matrix cells isolated from the murine forebrain vesicle (submitted).

Discussion

ROBERT BRENT (Philadelphia): Jack, that was one of the most comprehensive, clear talks that I have ever heard you give. It was really superb. I think you didn't tell the truth when you said you didn't prepare. Just a couple of comments; I think you ought to clarify the fact that even the early workers like Plummer and Yamazaki had a fifty percent mental retardation risk in the one gray area. So that's not new information. I think that new information is in the refinement at the lower dosages in the neuropathological findings. The second thing is that I think this is a time when animal data can be helpful. You know you can't do these experiments that you're talking about in the human. You mentioned Brizzee's work and I think that probably fractionation and protraction of radiation is even more dramatic in the embryo than it is in the adult with regard to the ability of the embryo to recuperate. If you remember his paper, one of his early papers where he had 1.5 Gray to the brain in the fetal stage and divided it, instead of one dose, into nine doses over a period of twelve hours and was unable to show any neuropathological effect with just a twelve hour fractionation. So I think those are important studies. Our own work in the area of neuro behavioral studies, which we've been extending over the last two or three years, indicates that you can get severe neuro behavioral effects in the rodent model in the ranges that you are talking about, the fifty to 1.5 Gray. When you get down to 0.2 Gray, they disappear. So I think that studies correlating neuropathology and behavior in the 0.01 to 0.2 Gray range in the animal models are going to be extremely helpful to try to clarify some of these points. You know we always say that the human is the best model but I think that these studies just aren't able to be done. Finally, you know the literature keeps repeating the fact that you don't get mental retardation before the eighth week. I don't think we have the data to say that. First of all, the LD_{50} for the embryo during organogenesis is very low. It's about 110 to 120 rads, for a 28 day human gestation, if you can use any of the mammalian data. This means that in Hiroshima and Nagasaki, there was a lot of embryonic loss. I don't know how many embryos survived with neuropathological effects because the LD_{50} was low. I think you could probably do a schematic of Hiroshima and Nagasaki, and find out how many women were twenty-first day to thirty-fifth day during the early period of organogenesis. I think you find there are very few who were pregnant at that stage. Then figure out how many of them could have gotten between seventy-five and hundred rad, and I think you'd find that there were very few. So, I don't think we can say anything about open neuro tubes, mental retardation during that period, except that you didn't see it.

WILLIAM SCHULL: I don't take exception to anything that Dr. Brent said. We have ourselves felt that if you take clinical retardation as we measured, we don't see it in the first eight weeks. That obviously doesn't mean that there could not have been severe damage not only to the brain but other organs as well. There was a heightened pregnancy loss as James Yamazaki showed in 1954. And there are, I think, embryological reasons and physiological reasons for believing that the damage to the brain stem, for example, would probably be incompatible with life if it was at all severe. So I think we do not see the effects, not because they may not have occurred, but because they didn't survive to be measured.

Experimental Radiobiology and Radiation Protection Studies

R. J. M. Fry
Oak Ridge National Laboratory
Oak Ridge, TN

Introduction

The estimates of risk of genetic effects in humans, by necessity, have been based on experimental animal data. As far as can be judged at this time the estimates are not inconsistent with the limited human findings. In contrast experimental animal data have played a very minor role in the estimation of risk of radiation-induced cancer in humans. The simple reason is that human data have been available and no one can argue against Alexander Pope's advice that

> "... the proper study of mankind is man, a being darkly wise and rudely great with too much knowledge for the skeptic side with too much weakness for the stoic's pride ..."

The risk of cancer rather than genetic effects are the salient late effects of concern. It is the risk of excess cancer that will largely govern the selection of radiation protection standards. At the same time as data from the human experience with radiation accumulates and therefore risk estimates improve it is becoming clear that there is information about radiation-induced late effects that is unlikely to be obtained from humans now or in the future. Data from animal experiments will have to be used to help in human risk estimates.

In the past the contribution made by animal and cellular studies to risk estimation of radiation-induced cancer has been in the study of mechanisms and establishing what Mole called "useful generalizations" (1) such as the influence of dose rate and radiation quality. Although the fact that the relative biological effectiveness (RBE) increases with linear energy transfer (LET) is a useful generalization, it only becomes useful in the estimation of human risks if a quantitative estimate of RBE for a particular radiation and endpoint can be made and extrapolated across species.

Perhaps it is timely to assess the role of experimental studies in providing information that is required for estimating radiation risks and setting protection standards. Timely, because human studies have not provided the data required to assess the effect of radiation quality or dose rate and fractionation sufficiently for radiation protection purposes.

The new dosimetry for atomic bomb survivors (2) has reduced greatly, if not eliminated, the possibility of deriving risk estimates for neutron-induced cancer. The neutron doses are now estimated to be considerably lower than the T65 dose estimates. There is little or no likelihood of obtaining risk estimation for cancer induction in humans by other high-LET radiations, such as heavy ions. The importance of heavy ion radiobiology is limited to radiation protection of a small group of people but for that group the information is of considerable importance.

Although the great majority of radiation exposures, to either worker or the general population are incurred at low dose rates (mGy per yr) or to very small dose fractions of high dose-rate exposures, it is not known quantitatively how the carcinogenic effect of such exposures compares with those of high dose-rate irradiation. The important biological aspect of the industrial and environmental time-exposure patterns is that not only is the dose incurred at a low dose rate but it is also protracted over a considerable time. When radiation exposures are protracted over years then it is not only a matter of dose rate that is important but also the change in susceptibility. Susceptibility for radiation-induced cancer decreases with age, at least for most cancers that have been adequately studied. The protraction of irradiation over a working lifespan of about 40 years will reduce the carcinogenic effect independently and in addition to the reduction of the effect of lowering the rate of exposure to radiation. There is some quantitative information about the lower risk of radiation-induced cancer in older populations (3) but none for the effect of dose rate or the combined effects of dose rate and the age-dependent changes in humans that together can be called protraction effects.

Dose-Response Relationships

(a) Low-LET radiation

Models for the dose-response relationships of cancer induction of radiation stem from the work of Lea and Catcheside (4) and Sax (5) on the induction of chromosome aberrations. How the deposition of energy leads to cellular effects has been given considerable attention but the linear and linear-quadratic models proposed by the early workers for the responses of high- and low-LET radiation, respectively, have held their popularity.

Models such as the dual radiation action formulation of Kellerer and Rossi (6) and that of Bond *et al.* (7) have related the characteristics of the deposition of energy at the subcellular and cellular level to the probability of the effect. Others such as Chadwick and Leenhouts (8) and Heartlein and Preston (9) have been concerned with the nature of the biological lesions that may lead to the effects and Goodhead has made estimates of the size of the targets involved (10). Moreover, none of these models are concerned specifically with carcinogenesis. Considering the complexity of carcinogenesis it is not surprising that there have been very few attempts to model the complete process of experimental radiation carcinogenesis. Vanderlaan *et al.* (11) have modeled radiation induction of skin cancer and Marshall and Groer (12) bone tumor induction.

In Fig. 1 is shown the widely accepted dose-response curve for cancer induction by low-LET radiation and its formal description. As is indicated by the notations on the figure there are still a lot of unanswered questions. The formula describing the curve does little justice to various dose- and time-dependent processes involved. Nor is the formulation necessarily appropriate for a process that is thought to have multiple stages. What is required is more modeling on the lines of Vanderlaan *et al.* (11), Marshall and Groer (12) and the more recent work of Moolgavkar (13). I will consider only one feature of the curve, namely, the response to low doses. The linear component of the initial part of such dose-response curves for the induction of chromosome aberrations is interpreted as representing single track events. It is clear that this cannot be the case for cells or even nuclei as it has been estimated that the doses at which nuclei are traversed on the average by one track are very much lower than at which there are any observations for the effects of low-LET radiation (10). Presumably the linear component is related to single-track events in the pertinent target and therefore the size of the target may be a determining factor. It can be seen from the schematic

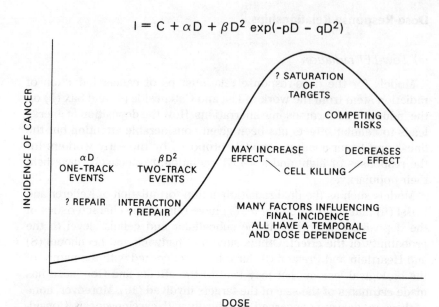

Fig. 1. Schematic of the incidence of cancer as a function of dose.

in Fig. 2 that the linear component extends over very different ranges of dose for the induced tumor responses in different tissues. The reasons for these differences have never been explained apart from saying, with an air of apparent knowledge, that differences are due to the differences in the mechanisms of different cancers—an answer that sounds good but means little.

The dose-response relationships based on a linear-quadratic model imply that the initial slope of a single dose-response curve should be equivalent to the slope of the response after low-dose rate exposures and multiple fractions each of which is a small dose. That aspect of the dose-response models has been tested experimentally only recently (14). The incidences of lung and mammary carcinomas in mice as a function of single, multiple and low dose rate exposures are shown in Fig. 3. In the case of lung carcinoma the linear component appears predominant over a dose range of about 0.0–0.2 Gy. If the linear-quadratic model is appropriate the slope of the response to multiple fractions of less than about 0.2 Gy should be the same as the initial slope of the single dose-response curve and the slope of the response to low dose-rate exposures, in this case 0.083 Gy/day. While the number of data points leaves a lot to be desired, the indication is that the response to 20 fractions of 0.1 Gy fits the model. The curve for the induction of breast cancer is similar

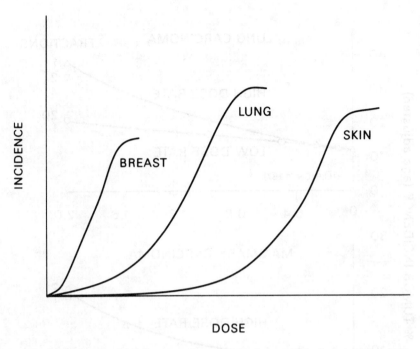

Fig. 2. Schematic of the dose-response curves for tumors of the breast, lung and skin in the mouse to illustrate the differences in the shapes of the curves, in particular the initial slopes.

in form but very different in detail. In this case a linear component is difficult to delineate experimentally, although a linear-quadratic fit to the dose-response curve is considered the best fit. Note that the dose scales for the dose-response curves for lung and breast cancer are different. In this case breast cancer reduction of the exposure dose rate also reduces the effect. However, in the case of fractionation the dose per fraction must be reduced to about 0.01 Gy per fraction to reduce the effect per unit dose to the level obtained with low dose-rate exposure (0.1 Gy/day). In these two tissues, although the dose-response curves are different, the effect per unit dose, with single doses in the range of the linear component, is the same as that with multiple fractions and low dose-rate irradiation.

The importance of the results is not only the confirmation of the linear-quadratic model but the results provide evidence that the carcinogenic effects of low doses, that are well nigh impossible to determine directly, can be determined using low dose-rate exposures with reasonable total doses. We need more data for various tissues obtained with

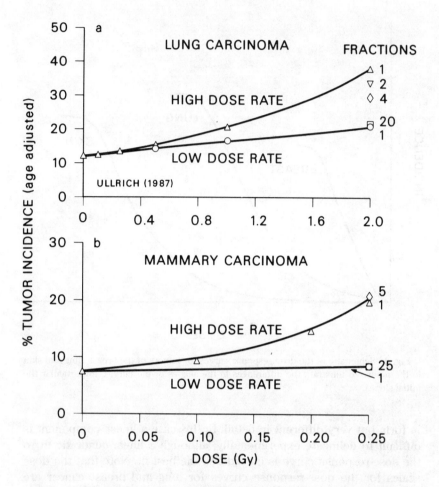

Fig. 3. Age-adjusted incidences of lung and mammary cancers as a function of high dose-rate (△–△) and low dose-rate (o–o) gamma radiation. The incidence after fractionated-dose regimens are indicated by symbols and the number of fractions on the right hand side of the plots. For example, the incidence of lung carcinoma after 4 fractions, 0.5 Gy/fraction and total dose of 2.0 Gy is shown by ◇4 in the upper panel. Data from Ullrich et al. (18).

exposures at low dose rates that will not only establish the effects of lowering dose rate but can be used to estimate the effects of small doses.

Nothing is easy in experimental carcinogenesis and low dose-rate and fractionation experiments are no exception. In order to obtain tumor incidences that are significantly in excess of control levels relatively high total doses are used in both low dose-rate and fractionation exper-

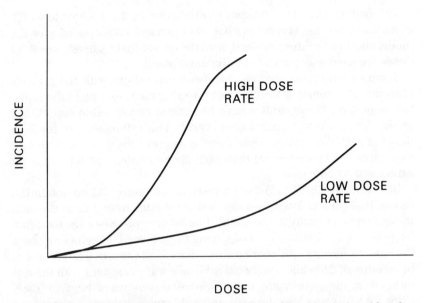

Fig. 4. Schematic of incidence as a function of dose delivered at high and low dose rates. The figure illustrates the finding that at high total doses of low dose-rate irradiation the incidence per unit dose increases.

iments. Above some total dose the effect per unit dose increases even at low dose rate as shown schematically in Fig. 4. This effect has also been noted in chemical carcinogenesis experiments. There is a possibility that repeated fractions or protracted exposure have what would be termed in the jargon of skin carcinogenesis a "promotion" effect. There are many factors involved in the expression of initial events of carcinogenesis. While the initial events are essential it is the factors influencing expression that determine the cancer rates. Less is known about the factors influencing expression than about initiation and little is known about how protracted exposures to carcinogenic agents affect these factors.

Experimentally, there are three approaches to the estimation of the effects of very low doses of low-LET radiation: 1) the direct determination of the initial slope of the single dose response, 2) the indirect estimation from data obtained at high doses and extrapolation to low doses using some model of the dose-response relationship, and 3) the estimation of the initial slope of the dose response obtained from the slope of the response to a range of doses delivered at very low dose rates.

The first method is exorbitantly expensive in the case of low-LET radiation, while the second method is dependent on the validity of the model chosen. The third method, with the caveat that relatively low total doses are used may prove to be the most practical.

Some experimental studies have been carried out with the primary objective of estimating the effect of lowering the dose rate of the radiation exposure. These studies have become of considerable importance as the data for human exposures used in risk estimation are for high doses at high dose rates. While there is a consensus that low dose-rate exposures are less effective than high dose-rate exposures there is no agreement by how much.

In 1977, UNSCEAR (15) used a reduction factor of 2.5 to obtain the risk of leukemia at low doses and low dose rates from a fit to the data for leukemia mortality in the atomic bomb survivors. Since the total risk for cancer was obtained by multiplying the adjusted leukemia risk by 5 the reduction factor was applied *pari passu* to all cancers. The reduction by a factor of 2.5 while somewhat arbitrary was consistent with the α/β ratio of a linear-quadratic model. Some eleven years later in 1988, UNSCEAR (3) was less decisive and only suggested that estimates of risk obtained from the high dose rate data should be reduced by a factor between 2 and 10! The misfortune of such uncertainty is that if the correct factor is 10 then the new risk estimates for an adult population (3) would be no greater than those estimates made in 1977 (15) but they would be greater if the reduction factor is really 2. Can experimental animal data resolve this dilemma?—not immediately from the reported data.

In NCRP Report 64 (16) all the available experimental data on dose-rate effects on cell killing, mutation, chromosome aberrations and cancer induction were examined. The consistency of the findings of a reduced effect with a reduction dose-rate exposure is compelling. The Dose Rate Effectiveness Factor (DREF) was chosen to express the numerical value of the reduction of the carcinogenic effect by reduction in dose rate and is approximately given by:

$$\text{DREF} = \frac{\text{The effect per unit dose at high-dose rate}}{\text{The effect per unit dose at low-dose rate}}$$

where the effects per unit dose are the linear regression coefficients obtained from data for the high and low dose-rate exposures.

This method provides a fairly crude approximation of the DREF for any particular dose. The linear regression of high dose rate must only be applied to data appropriately adjusted for competing risks. As has been discussed above the effect per unit dose may increase at high total doses even when the exposure is at a low dose rate. In the estimates of the effect per unit dose reported for the induction of myeloid leukemia and thymic lymphoma, data which may have reflected this total dose dependence was used (16). There is a need for some new experiments using low dose-rate exposures for which the design takes into account what we have learned over the years.

The error introduced by the use of a single DREF for the induction of various types of tumors has never been assessed. The last point that might be made about dose-rate reduction factors is that even though the response to high doses may be linear, lowering the dose rate will result in a lowering of the carcinogenic effect. This has become an important point because the most recent dose-response curves for radiation-induced cancers appear to be more linear than previously thought.

(b) High-LET Radiation

What is surprising about high-LET radiation dose-response relationships is that there has been no formulation that describes the response over a reasonable dose range. There is an assumption that the initial slope of the dose-response curves for neutrons are linear but without any unequivocal experimental data to support the contention. In fact some of the analyses of neutron dose-response data have used a power function. Carnes et al. (17) have discussed the various models used in the analysis of neutron-induced life shortening. The general conclusion is that no single equation that has been applied describes adequately the dose-response curves for neutrons.

The data for the incidence of Harderian gland tumors as a function of dose of Argon-40 can be fitted with a linear response in the dose range of 0.0 and 0.2 Gy with considerable confidence (Fig. 5). It is the response seen above 0.2 Gy in Figure 5 that has not yet been formally described for tumor dose-response curves for high-LET radiation. It has been suggested that the bending over of the curves that is also seen with neutrons and low-LET radiation is caused by cell killing. In the case of mammary tumors in BALB/c mice (18) the neutron dose-response curve begins to bend over at 0.05 Gy, a dose level at which it would be difficult to account for the significant change in slope of the curve on the basis of cell killing. The bending over at such low doses makes it difficult to establish experimentally the initial slope of the dose-response curves. The determination of the initial slope is important for both direct estimates of risk for neutron effects and also for the calculation of RBEs.

The first late effects experiment carried out with JANUS reactor at Argonne National Laboratory (19) was designed to examine whether

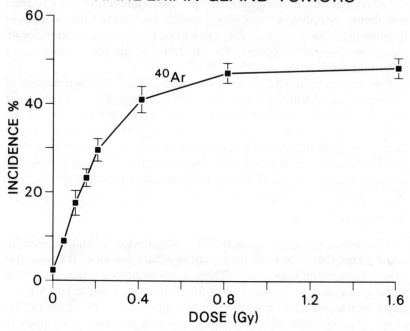

Fig. 5. Incidence of Harderian gland tumors as a function of dose of 570 MeV/amu Argon-40.

additivity held for fission neutrons with life shortening and tumor induction as the endpoint. The early results indicated that the effects on life shortening and tumor induction were greater with fractionated exposures than with single doses (20). Later it became clear that the increased effect with fractionation, seen at doses as low as 60 cGy did not occur or could not be demonstrated at doses below 20 cGy (see 17 and 21 for review of the various studies).

Some years later, Hill et al. (22) reported an increased effect of the same JANUS fission-spectrum neutrons on neoplastic transformation of C3HL10T1/2 cells with low dose-rate exposures to neutrons. The increased effect occurred at low total doses and became known as the inverse dose-rate effect. Similar enhancement was found with fractionated neutron exposures. These results have stimulated great interest but no satisfactory explanation. There are conflicting reports from both *in vitro* and *in vivo* studies about inverse dose rate and fractionation effects. The resolution of these conflicts will be interesting from a radiobiological point of view but there is little evidence that an inverse dose-rate effect has any great impact on estimates of risk that are of consequence for radiation protection.

RBE

The hope that estimates of risk of neutron-induced cancer would come from survivors at Hiroshima has been dashed by the new dosimetry (2). For the immediate future the choice of Q for neutrons will have to be based on data for the induction of chromosome abberations in human cells *in vitro* and tumors in experimental animals. The RBE of interest to radiation protection is the maximum RBE (RBE_{max}) which is the only meaningful single value and is estimated as the ratio of the initial slopes of the responses for neutrons and the reference radiation. The historical reference radiation was often x rays. However, any of the data that meet the criteria required for RBE determinations have been obtained using gamma rays as the reference radiation. The introduction of the use of RBE_{max} has eliminated some of the problems of this indirect method of assessment but not all. Any particular RBE is more dependent on the response to the reference radiation at the very low doses involved than to the response to the neutrons; an undesirable situation. In Figure 6 data for the induction of myeloid leukemia by gamma rays and neutrons (23) is shown. Linear dose responses are consistent with the data for both the neutron and gamma-ray exposures. However, the data for the lowest gamma doses indicate that the initial slope may be much less than estimated from the linear fit and the RBE could be very high. Clearly, there is need to determine the risks from neutron effects directly from the neutron dose-response curves and not rely on an indirect factor such as RBE. In order to use such estimates of high-LET radiation effects in radiation protection there has to be an accepted method of extrapolating the estimates across species.

Extrapolation of risk estimates across species

Brown *et al.* has reviewed methods of extrapolation of risks across species that have been suggested for the carcinogenic effects of chemicals and radiation (24). The advantage of developing methods of extrapolation across species of radiation effects is that the methods can be tested because data for both humans and experimental animals are available.

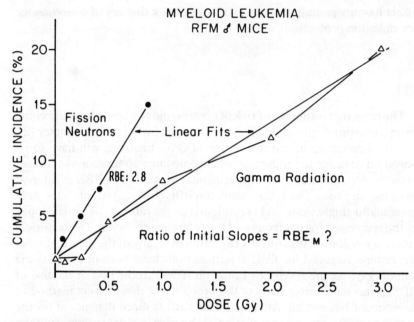

Fig. 6. Cumulative incidence of myeloid leukemia as a function of dose (Gy) of fission neutrons and gamma rays.

Of the methods suggested for extrapolating radiation risks across species three appear to have promise. First, the toxicity ratio suggested for use with radionuclides by Mays et al. (25). Second, the Bayesian approach suggested by Du Mouchel and Groer (26) and third, the relative risk method described by Storer et al. (27). The basis of the relative risk method is that susceptibility is influenced by the natural incidence and when the relative risk model is used to estimate risk of cancer the marked differences in the natural incidence of various cancers between species become unimportant. The comparative estimates of relative risk of lung and breast cancer and leukemia in mice and humans at 1 Gy of low-LET radiation are shown in Table 1. The results are encouraging but more studies are required.

TABLE 1—*Relative Risk (at 1.0 Gy) for Cancer in Humans and Mice*

Cancer	Mice	Atomic Bomb Survivors
Lung	1.8 (1.5–2.2)	1.7 (1.4–2.0)
Breast	2.0 (1.6–2.6)	1.9 (1.4–2.5)
Leukemia	5.4	6.3 (4.9–8.1)

It will be particularly important to obtain risks of cancer mortality at low dose rates in mice that can be used for extrapolation to humans when methods of extrapolation are validated.

Summary

In the past the contribution of experimental animal data to the risk estimates for external radiation carcinogenesis has been limited. In the future the contribution will have to be greater if the influence of dose rate and radiation quality on radiation-induced cancer are to be assessed. It is important that accepted methods of extrapolation of risk estimates across species are developed and there appear to be some methods with promise. The study of mechanisms of radiation carcinogenesis are central to experimentalists' studies and the understanding of the mechanisms will help to formulate better models, for example, of dose-response relationships but such studies will not solve all the problems of risk estimation.

References

1. Mole, R.H. "First Esler Comitato Nazionale Energia Nucleare. Late effects of exposure to ionizing radiation in animal experimentation and human protection," pages 179–199 in *Proceedings of First European Symposium on Late Effects of Radiation* (CNEN, Rome) (1965).
2. RERF. Radiation Effects Research Foundation. *U.S.–Japan Joint Reassessment of Atomic Bomb Radiation Dosimetry in Hiroshima and Nagasaki, Vol. 1 Final Report,* Roesch, W. C., Ed. (Hiroshima Radiation Effects Research Foundation, Hiroshima, Japan) (1987).
3. UNSCEAR. United Nations Scientific Committee on the Effects of Atomic Bomb Radiation. *Sources, Effects and Risks of Ionizing Radiation,* Report to the General Assembly, with Annexes (United Nations, New York) (1988).
4. Lea, D.E. and Catcheside, D.G. "The mechanism of the induction by radiation of chromosome aberrations in *Tradescantia,*" J. Genetics **44,** 216 (1942).
5. Sax, K. "Chromosome aberrations induced by x-rays," Genetics **23,** 494 (1938).
6. Kellerer, A.M. and Rossi, H.H. "A generalized formula of dual radiation action," Radiat. Res. **75,** 471 (1978).
7. Bond, V.P., Varma, M.N., Sondhaus, C.H. and Feinendegen, L.E. "An alternative to absorbed dose, quality and RBE at low exposures," Radiat. Res. **104,** 552 (1985).

8. Chadwick, K.H. and Leenhouts, H.P. "A molecular theory of cell survival," Med. Biol. **18,** 78 (1973).
9. Heartlein, M.W. and Preston, R.J. "An explanation of interspecific differences in sensitivity to x-ray induced chromosome aberrations and a consideration of dose-response curves," Mutat. Res. **150,** 299 (1985).
10. Goodhead, D.T. "Deductions from cellular studies of inactivation, mutagenesis and transformation," pages 309–385 in *Radiation Carcinogenesis: Epidemiology and Biological Significance*, Boice, J.D., Jr. and Fraumeni, J.F., Jr. (Raven Press, New York) (1984).
11. Vanderlaan, M., Burns, F.J. and Albert, R.E. "A model describing the effects of dose and dose rate on tumor induction in rat skin," pages 253–263 in *Biological and Environmental Effects of Low-Level Radiation, Vol 2*, IAEA/ST1/PUB/409 (International Atomic Energy Agency, Vienna) (1976).
12. Marshall, J.H. and Groer, P.G. "A theory of the induction of bone cancer by alpha radiation," Radiat. Res. **71,** 149 (1977).
13. Moolgavkar, S.H., Dewanji, A. and Venzon, D.J. "A stochastic two-stage model for cancer risk assessment. I. The hazard function and probability of tumor," Risk Analysis **8,** 383 (1988).
14. Ullrich, R.L., Jernigan, M.C., Satterfield, L.C. and Bowles, N.D. "Radiation carcinogenesis: time-dose relationships," Radiat. Res. **111,** 179 (1987).
15. UNSCEAR. United Nations Scientific Committee on the Effects of Atomic Bomb Radiation. Annex 1 pages 565–654 in *Sources and Effects of Ionizing Radiation* (United Nations, New York) (1977).
16. NCRP. National Council on Radiation Protection and Measurements. *Influence of Dose and its Distribution in Time and Dose-Response Relationships for Low-LET Radiations*, NCRP Report 64 (National Council on Radiation Protection and Measurements, Bethesda, Maryland) (1984).
17. Carnes, B.A., Grahn, D. and Thompson, J.F. "Dose-response modeling of life shortening in a retrospective analysis of the combined data for the JANUS program at Argonne National Laboratory," Radiat. Res. **119,** 39–56, (1989).
18. Ullrich, R.L. "Tumor induction in BALB/c mice after fractionated or protracted exposures to fission-spectrum neutrons," Radiat. Res. **97,** 587 (1984).
19. Grahn, D., Ainsworth, E.J., Williamson, F.S. and Fry, R.J.M. "A program to study fission neutron-induced chronic injury in cells, tissues and animal populations utilizing the JANUS reactor of the Argonne National Laboratory," pages 211–228 in *Radiobiological Applications of Neutron Irradiation*, ST1/PUB/325 (International Atomic Energy Agency, Vienna) (1972).
20. Ainsworth, E.J., Fry, R.J.M., Williamson, F.S., Brennan, P.C., Stearner, S.P., Yang, V.V., Crouse, D.A., Rust, J.H. and Borak, T.B. "Dose effect relationships for life shortening tumorigenesis and systemic injuries in mice irradiated with fission neutron or ^{60}Co gamma radiation," pages 1143–1151 in *Proceedings of IRPA IVth International Congress*, Paris (International Radiation Protection Association, Fontenay aux Roses, France) (1977).
21. Thompson, J.F. and Grahn, D. "Life shortening in mice exposed to fission neutrons and gamma rays," VIII Radiat. Res. **118,** 151–160 (1989).

22. Hill, C.K., Buonaguro, F.M., Mijers, C.P., Han, A. and Elkind, M.M. "Fission-spectrum neutrons at reduced dose rates enhance neoplastic transformation," Nature **298,** 67 (1982).
23. Ullrich, R.L. and Preston, R.J. "Myeloid leukemia in male RFM mice following irradiation with fission spectrum neutrons or γ rays," Radiat. Res. **109,** 165 (1987).
24. Brown, S.L., Brett, S.M., Gough, M., Rodricks, J.V., Tardiff, R.G. and Turnbull, D. "Review of interspecies risk comparisons," Regulatory Toxicology and Pharmacology **8,** 191 (1988).
25. Mays, C.W., Taylor, G.N. and Lloyd, R.D. "Toxicity ratios: their use and abuse in predicting the risk from induced cancer," in *Life-Span Radiation Effects Studies in Animals: What Can They Tell Us? Proceedings of the 22nd Hanford Life Sciences Symposium* Richland, Washington, 1983, CONF-830951 (National Technical Information Services, Springfield, Virginia) (1986).
26. Du Mouchel, W. and Groer, P.G. "Bayesian methodology for scaling radiation studies from animals to man," Health Phys. (in press).
27. Storer, J.B., Mitchell, T.B. and Fry, R.J.M. "Extrapolation of the relative risk of radiogenic neoplasms across mouse strains and to man," Radiat. Res. **114,** 331 (1988).

Discussion

JOHN BATEMAN (formerly from Brookhaven Laboratory): I would like to compliment Dr. Fry on a very clear and informative presentation. Particularly bringing out the mechanism likely for the high RBE of neutrons at low doses and the decline as with increasing dose. Our own work with lens opacification in the mouse yielded, with monoenergetic neutrons, RBE's well over a hundred for twenty millirads. Whereas at 40 rads the RBE was 10. Thank you again.

R.J. MICHAEL FRY: I am sorry that time has not permitted the coverage of a great deal of work of the people whom I see in the audience.

MICHAEL SIMIC (NIST): If you accept me as a young investigator, I'm willing to write that essay. Well the formula is you have to do molecular studies with animals not with cells. And unfortunately there is very little work of that kind. And you have to understand that all kinds of indigenous processes that are going on without any radiation *e.g.*, formation of free radicals, formation of the DNA damage and measurement of these biomarkers as a function of age, metabolic rates, and dose.

HARALD ROSSI: I think that there is quite a good argument to be made for the approach of relative risk which you mentioned. A reason being that very remarkably the RBE, for instance of neutrons, reaches a maximum, depending on neutron energy, in the same way for practically any higher organism we know. Furthermore, the RBE depends on dose the same in practically any organisms, higher organisms, we know. Which clearly indicates that we're dealing with some basic lesions which are the same, probably in DNA. So the logical conclusion to draw from that is that for a number of lesions, dependence of effect on lesions varies, but that the basic lesion production is a major point whether by radiation or otherwise.

Genetic Risk Estimates: Of Mice and Men

S. Abrahamson
Radiation Effects Research Foundation
Hiroshima, Japan

It is a pleasure and distinct honor to be part of the sixtieth anniversary program of NCRP. We celebrate an important institution whose body of work has achieved international respect. I have been asked to provide both an overview and critique of the last thirty years of genetic risk estimates, in approximately a twenty minute oral presentation.

Actually the overview should begin at the very foundation of the field of radiation genetics when the independent discoverers H.J. Muller and L.J. Stadler came to opposite conclusions regarding the nature of the radiation induced gene mutation event (see discussion in NCRP Report 64) (1). Muller was of the view that radiation induced mutations were intragenic changes capable of reversion to normal. Stadler on the other hand believed (based on his studies with maize and other plants) that these induced mutations were small chromosome aberrations, primarily deletions (intra or intergenic in size). While the controversy still is waged in different research laboratories, in my view the evidence appears to overwhelmingly favor the Stadler view that radiation induced mutations are primarily of the deletional or other chromosomal aberration type. This conclusion is based on the molecular analyses of induced mutations at selected gene loci in eukaryotes. The first corollary of immediate interest is that the dose response relationships for mutation induction should parallel that of chromosome aberrations in the germ cell stages under investigation and for equivalent doses.

In one sense it has been fortunate that the Drosophila-derived concept of the linearity of mutation induction with dose, preceded the pioneering efforts of the 1st BEAR, (2) and U.N. reports on the subject, (3) because it set the tone of prudence that has been followed in all subsequent analyses. Since then the last three decades of genetic risk estimation have been dominated by mouse-derived values, again fortunately a very radiosensitive set of specific loci have served as the main source of these estimates for most of this period but for spermatogonial cells only. It was with this system that Russell *et al.* (4) were first able to demonstrate a dose rate effect of chronic gamma exposure relative to acute xrays.

The major basis for risk estimation established by these committees involved the use of the Doubling Dose Concept. As originally developed the procedure entailed an estimate of the total current incidence of genetic disease in the population of about 2–4% and the dose required to be given to each generation which, at equilibrium, would lead to a doubling of this incidence. Based on acute doses to mouse spermatogonia this value was estimated to range between 10–70 rad with a mean value of about 40 rad. Explicit in their analysis was the statement that the derived doubling dose would be correct only if the assumption of linearity with dose was true. Note—that two elements were involved: the whole of the genetic disease base and the linearity assumption of induced mutation.

Within a few short years, the discovery of the dose rate effect for induced mutations at these seven loci established a doubling dose that was more appropriate for assessing the population risk from low doses and low dose rates of irradiation. (The more usual population exposure condition.) Furthermore it was later to become apparent to many radiobiologists that the linearity assumption of mutation induction for acute high doses of radiation was in conflict with the low dose-rate observations where linearity did occur over a wide range of doses (see for example (5) and (1)). I will return to this point later in the discussion because it may help to resolve an existing controversy. In addition, the genetics risk estimating community became more selective in the use of doubling dose for genetic conditions, primarily applying the procedure to only single gene Mendelian dominants and X-linked disorders and to only a select portion of the large bulk of diseases—the irregularly inherited congenital and constitutional diseases. The mutation component concept was introduced to delimit that fraction of the incidence that is directly proportional to the mutation rate.

An examination of the next few Tables will elaborate the major changes in risk estimating procedures after the early BEAR (2) and UNSCEAR (3) attempts. With few exceptions, refinements in the radiation aspects have involved fine tuning the calculations; introducing a better understanding of cytogenetic events, *e.g.* chromosome aberrations resulting from changes in number or structure. Human cytogenetics really emerged as a field in the 1960's, well after the publication of the first reports.

TABLE 1—*Genetic Effects—Population Exposure*
1 Rem (average)

Committee	Current Incidence Mendelian Traits†	Doubling Dose	No. Cases/ 10⁶ Liveborn 1st Generation
BEAR 1956	2%	40	50
BEIR I 1972	2%	20–200	10–100
BEIR III 1980	1.7%	50–250	10–40
BEIR V 1989		About same as BEIR III	

†Includes, Dominants, X-linked and chromosome aberrations.

A second approach using mouse dominant skeletal mutations was introduced in the 1980 BEIR report (6) and with its many, many, assumption provide a similar risk estimate to those shown here. Clearly what is obvious from Tables 1 and 2—BEIR (7) and UNSCEAR (8) respectively is that neither the frequency of Mendelian diseases nor the risk estimates have changed dramatically in this period. It should be obvious given the range of error in the incidence value it would be impossible to demonstrate an effect on a population of a 1 rem exposure to one million parents.

However, if we look at the class of diseases known as complexly inherited multifactorial diseases the evaluation is in some major ways quite different Tables 3 and 4. From 1956 to 1980 there was about a three fold change in the estimated natural frequency of these diseases from 3% to about 10%, and great uncertainty in calculating the risk. From 1980 to now, the incidence has inflated from seven to over ten fold based on studies carried out in Hungary (9). In other words these committees

TABLE 2—*Genetic Effects—Population Exposure*
1 Rem (average)

UNSCEAR Committee	Current Incidence Mendelian Traits	Doubling Dose	No. Cases/ 10⁶ Liveborn 1st Generation	
		Acute		
1958	1%	10–100	20–200	
1966	1%	70	40	
		Chronic		
1972	1%	100	6–15	
1977	1.5	100	20+38	(26)†
1982	1.6	100	18	(20)
1986–88	1.3	100	18	(25)

†Numbers in ( ) derived by "Direct" method rather than Doubling Dose.

TABLE 3—*Current Incidence of Congenital Anomalies and Multifactorial Diseases*

Committee	Incidence/ 10^6 Liveborn	No. Cases per rem per 10^6 Liveborn 1st Generation
BEAR 1956	3%	?
BEIR I 1972	4%	1–100
BEIR III 1980	10%	2– 90
BEIR V 1989	>50%	?
UNSCEAR 1988	66%	(10–500) not estimated

now suggest on average, every liveborn individual may suffer from at least one serious genetic disease in their lifetime. The BEIR committee apparently includes heart disease and cancer within this body of disorders for which a genetic basis might exist. These committees have put themselves "out on a limb" but not in a really perilous position because neither committee is now prepared to make a quantitative risk estimate. By their previous estimating procedures it is clear that the risk could be much larger than all other types or negligibly small. Basic research in this area has been very limited.

This has always been an area of great uncertainty with respect to both the mutation component (M.C.) and a reasonable estimate of equilibrium time. BEIR 1972 (10) suggested that the M.C. probably ranged between 5 and 50% and the equilibrium time was probably not less than 10 generations. UNSCEAR adopted BEIR's 5% value and estimate of equilibrium time in the early 80's but rejected them in 1986 about the time the study in Hungary (9) was showing unexpected high rates. It is important to recognize that in this study the great increase in such complexly determined diseases occurs after age 20 through to about age 60. Mouse experiments are as yet incapable of this type of analysis.

I would like to now return to the issue of doubling dose for acute exposure. Neel and colleagues for the past four decades have been attempting to determine if the F_1 offspring of the A-bomb survivors (the

TABLE 4—*Current Incidence of Congenital Anomalies and Multifactorial Diseases*

UNSCEAR Committee	Incidence/ 10^6 Liveborn	No. Cases per 0.01 Gy per 10^6 Liveborn 1st Generation
1958–1972	2–3%	not calculated
1977–1982	9%	5†
1986–1988	66%	not calculated

†Calculation was based on a Mutational Component (M.C.) of 5%; BEIR employed M.C. range of 5–50%.

same group from whom cancer risks have been derived) show an increased incidence of genetic disease, namely untoward pregnancy outcomes, stillbirths, neonatal deaths, congenital anomalies balanced chromosome aberration, mortality to age 26, electrophoretic mutations and other endpoints. The same F_1 group or subsets have been followed through this period. Remember the parental exposure was to an acute dose with a mean of about 0.50 Gy, approximately the suggested doubling dose estimated from mouse high dose acute experiments. The regression analyses of the human data on the other hand would place the human doubling dose at about 1.50–2 Gy for acute doses and possibly at 2.5–4 Gy for low doses or dose rates, Neel et al. (16) and in press. The confidence limits on these values are quite large. The authors have originally suggested a 4–5 fold difference in radiosensitivities between mice and men. Dr. Neel and I have been involved in an eight month intensive dialogue on these issues. He has convinced me that as in the case of cancer relative risks, both the radioinsensitive and radiosensitive endpoints should be included in determining a human doubling dose. I may have convinced him that mouse and man are not as different as he suggests for the following reasons. Figure 1 shows the mouse doubling-dose estimates based on the usual linear extrapolation from acute doses, the low dose rate estimate and my estimate for the more likely linear-quadratic doubling dose. Note the linear extrapolation is approximately ½ that of the linear quadratic, which in turn is about 70% of the well established chronic exposure value. But also remember that this is for the most sensitive mutation endpoint studied in the mouse, the seven specific locus test. The lowest tested acute dose was 3.0 Gy (much larger than most of the A-bomb parental doses). I have attempted to estimate the doubling doses for a majority of other mouse endpoints usually studied at doses of 6 Gy or larger. Figure 2 shows the doubling doses extrapolated by linear methods range from 0.30 to 1.70 Gy. Using linear-quadratic extrapolation estimates most of them fall into the range estimated with the human data, namely about 0.70–2.50. The extrapolations were prepared from the summarized data presented in UNSCEAR (8) or their earlier reports. In each case I had to estimate the α term by employing conventional dose rate reduction factors based on the acute value and the endpoint studied. The β term of the linear quadratic was based on other studies, taken as 10^{-2} of the α term. Our own extensive Drosophila gonial studies, Abrahamson and Meyer (17) on some 800 recessive lethal loci tested at eleven doses, fit a linear-quadratic response. The doubling dose for Drosophila was 5 Gy, which could well serve as the upper bound estimate of the mammalian doubling dose range.

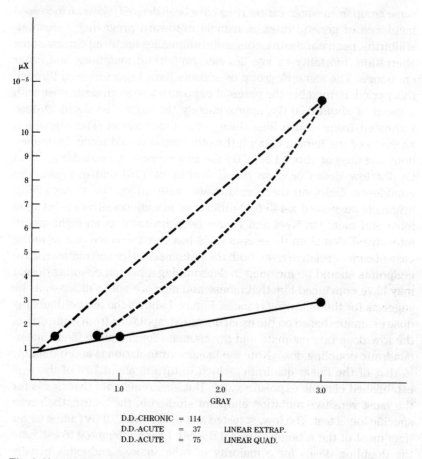

Fig. 1 Mouse—Spermatogonia

D.D.-CHRONIC = 114
D.D.-ACUTE = 37   LINEAR EXTRAP.
D.D.-ACUTE = 75   LINEAR QUAD.

At this point it may turn out that for a range of different endpoints in both mouse and man the results are not in real disagreement and if so it may well be that we are entering the new era of risk estimates wherein human studies can substitute for the mouse as its own surrogate.

Given the enormity of the genetic disease incidence projected for complexly inherited traits by BEIR (1989) (7) and UNSCEAR (1988), (8) mainly occurring after maturation and with the F_1 survivors now age 40 and younger it seems to me to be tragic to forego a major clinical analyses of the F_1 and their control over the next few years. No other cohort has been as thoroughly studied, 16,000 F_1 cytogenetically analysed, over 20,000 F_1 examined for electrophoretic variants. Long before we can

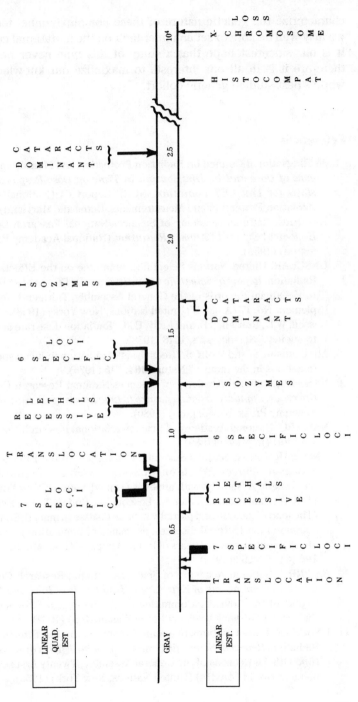

Fig. 2 Mouse Doubling Doses for Acute High Doses to Spermatogonia

characterize the genetic nature of these complexly inherited diseases we may well be able to set an upper limit on the mutational component. It is our sincerest hope that a study of this type never be repeated, therefore it is in all our interests to maximize our knowledge of the world's best studied genetic cohort.

References

1. NCRP. National Council on Radiation Protection and Measurements. *Influence of Dose and its Distribution in Time on Dose-Response Relationships for Low-LET Radiation*, NCRP Report 64 (National Council on Radiation Protection and Measurements, Bethesda, Maryland) (1980).
2. NAS/NRC. National Academy of Sciences/National Research Council. *The Biological Effects of Atomic Radiation* (National Academy Press, Washington) (1956).
3. UNSCEAR. United Nations Scientific Committee on the Effects of Atomic Radiation. *Report of Scientific Committee on the Effects of Atomic Radiation*, Official Records of the General Assembly, Thirteenth session, Supplement No. 17 (A/3838) (United Nations, New York) (1958).
4. Russell, W.L., Russell, L.B. and Kelly, E.M. "Radiation dose rate and mutation frequency," Science **128**, 1546 (1958).
5. Abrahamson, S. and Wolff, S. "Reanalysis of radiation-induced specific locus mutations in the mouse," Nature **264,** 715 (1976).
6. NAS/NRC. National Academy of Sciences/National Research Council. *The Effects on Populations of Exposure to Low Levels of Radiation* (National Academy Press, Washington) (1980).
7. NAS/NRC. National Academy of Sciences/National Research Council. BEIR V Report (in press) (1989).
8. UNSCEAR. United Nations Scientific Committee on the Effects of Atomic Radiation. *Sources, Effects and Risks of Ionizing Radiation*, Report to the General Assembly, with annexes (United Nations, New York) (1988).
9. Czeizel, A., Sankaranarayanan, K., Losconi, A., Rudas, T. and Keresztes, M. "The load of genetic and partially genetic disease in man. II. Some selected common multifactorial diseases: estimates of population prevalence and of detriment in terms of years of lost and impaired life," Mutation Research **196** (3) 259–292 (1989).
10. NAS/NRC. National Academy of Sciences/National Research Council. *The Effect on Populations of Exposure to Low Levels of Ionizing Radiation*. Report of the Advisory Committee on the Biological Effects of Ionizing Radiation (National Academy Press, Washington) (1972).
11. UNSCEAR. United Nations Scientific Committee on the Effects of Atomic Radiation. *Report of Scientific Committee on the Effects of Atomic Radiation*, Official Records of the General Assembly, Twenty-first session, Supplement No. 14 (A/6314) (United Nations, New York) (1966).

12. UNSCEAR. United Nations Scientific Committee on the Effects of Atomic Radiation. *Ionizing Radiation: Levels and Effects*, United Nations sales publication E.72.IX.17 and 18 (United Nations, New York) (1972).
13. UNSCEAR. United Nations Scientific Committee on the Effects of Atomic Radiation. *Sources and Effects of Ionizing Radiation*, Report to the General Assembly, with annexes, United Nations sales publication E.77.IX.1 (United Nations, New York) (1977).
14. UNSCEAR. United Nations Scientific Committee on the Effects of Atomic Radiation. *Ionizing Radiation: Sources and Biological Effects*. Report to the General Assembly, with annexes, United Nations sales publication E.82.IX.8 (United Nations, New York) (1982).
15. UNSCEAR. United Nations Scientific Committee on the Effects of Atomic Radiation. *Genetic and Somatic Effects of Ionizing Radiation*. Report to the General Assembly, with annexes, United Nations sales publication E.86.IX.9 (United Nations, New York) (1986).
16. Neel, J.V., Schull, W.J., Awa, A.A., Satoh, C., Otake, M., Kato, H. and Yoshimoto, Y. "Implications of the Hiroshima-Nagasaki genetic studies for the estimation of the humans "doubling dose" of radiation," Genetics Congress, Toronto (in press).
17. Abrahamson, S. and Meyer, H.U. "Quadratic analysis for the induction of recessive lethal mutation in drosophila oogonia by x-irradiation," page 9 in *Biological and Environmental Effects of Low Level Radiation.* 1, IAEA/STI/PUB/409 (International Atomic Energy Agency, Vienna) (1976).

Scientific Session

Perception

Leonard Sagan
Chairman

Perception of Risk from Radiation

Paul Slovic
Decision Research
and University of Oregon
Eugene, OR

In March, 1979, I had the privilege of addressing the 15th Annual Meeting of the NCRP on the topic of "Perception and Acceptance of Risk from Nuclear Power." When Warren Sinclair asked if I would provide an update on the topic of perception, broadening the scope to include radiation in general, I eagerly accepted. The topic is important, and the past 10 years have been very eventful. We have learned much about risk perceptions, though not enough.

When I say that the decade has been eventful, I mean *eventful.* Exactly at the time of the 1979 meeting (in mid-March), the movie titled *The China Syndrome* had its premiere. Two weeks later, events at Three Mile Island made the movie appear prophetic. Succeeding years have brought us Chernobyl and other major technological disasters, most notably Bhopal and the Challenger accident. The public has drawn a common message from these accidents—that technology is unsafe, that expertise is inadequate, and that government and industry cannot be trusted to manage hazards safely. These dramatic accidents have been accompanied and reinforced by numerous chronic problems involving radiation, such as the discovery of significant radon concentrations in many homes, the continuing battles over the siting of facilities to monitor and store nuclear wastes, and the disclosures of serious environmental contamination emanating from nuclear weapons facilities (at Hanford, Fernald, Rocky Flats, and Savannah River).

Dr. Sinclair asked me: "Has there been progress?" My answer is that there has been *change* in perceptions and *change* in our understanding of perceptions and their implications—but there has been little or no progress in our ability, as a society, to think calmly and constructively about the management of radiological technologies. This troublesome state of affairs is not unique to radiation technologies—we find that a parallel situation has emerged with regard to perception and management of many chemical technologies.

Perhaps the most important generalization to be drawn from risk-perception research is that there is no *uniform* or consistent perception of radiation problems. This is what makes this topic so fascinating to study from the standpoint of perception and acceptance of risks. Public perception and response is determined by the context in which radiation is used—and the very different reactions to different uses provide insight into the nature and causes of perception.

A second generalization, and a disturbing one, is that in every context of use, with the possible exception of nuclear weapons, public perceptions of radiation risk differ from the assessments of experts on radiation and its effects. Thus, I would say that the knowledge and opinions held by the members of the NCRP are generally not consistent with the views of most members of the public. In some cases, members of the public see far greater risks associated with a radiation technology than do technical experts—in others the public is much less concerned than the experts believe they should be. It is clear that better information and education about radiation and its consequences is needed. With the exception of studies that have designed brochures to help people understand their risk from radon, there has been little effort or progress made on the communication side.

There is a particularly urgent need to develop plans and materials for communicating with the public in the event of a radiological disaster. This point is driven home by the difficulties observed in Europe after Chernobyl, and in the chaos and disruption that reigned in Goiania, Brazil after two scavengers unwittingly sawed open a capsule containing cesium that had been used for cancer therapy.

This, in essence, is my story. During the remainder of this paper I shall attempt to fill in the details by highlighting some key results and conclusions pertaining to:

1. the nature of risk perceptions
2. the impacts of perceptions
3. the need for communication about radiological hazards.

The Psychometric Paradigm

One broad strategy for studying perceived risk is to develop a taxonomy for hazards that can be used to understand and predict responses to their risks. A taxonomic scheme might explain, for example, people's extreme aversion to some hazards, their indifference to others, and the discrepancies between these reactions and experts' opinions. The most common approach to this goal has employed the *psychometric paradigm* (1) which uses psychophysical scaling and multivariate analysis techniques to produce quantitative representations or "cognitive maps" of risk attitudes and perceptions. Within the psychometric paradigm, people make quantitative judgments about the current and desired riskiness of diverse hazards and the desired level of regulation of each. These judgments are then related to judgments about other properties, such as the hazard's status on characteristics that have been hypothesized to account for risk perceptions and attitudes (*e.g.*, voluntariness, dread, knowledge, controllability) and the benefits that each hazard provides to individuals and to society.

Results from these studies have shown that perceived risk is quantifiable and predictable. Psychometric techniques seem well suited for identifying similarities and differences among groups with regard to risk perceptions and attitudes (see Table 1). They have also shown that the concept "risk" means different things to different people. When experts judge risk, their responses correlate highly with technical estimates of annual fatalities. Lay people can assess annual fatalities if they are asked to (and produce estimates somewhat like the technical estimates). However, their judgments of "risk" incorporate other factors as well (*e.g.*, catastrophic potential, threat to future generations) and, as a result, tend to differ from their own (and experts') estimates of annual fatalities.

Factor-analytic representations. Many of the risk characteristics are highly correlated with each other, across a wide range of hazards. For example, hazards rated as voluntary tend also to be rated as controllable and well-known; hazards that appear to have catastrophic potential also tend to be seen as having fatal consequences, and so on. Investigation of these interrelationships by means of factor analysis has shown that the broader domain of characteristics can be condensed to a small set of higher-order characteristics or factors.

The factor space presented in Figure 1 is based on one of our early studies of college students in Oregon.[1] Factor 1, labeled "Dread Risk," is defined at its high (right-hand) end by perceived lack of control, dread, catastrophic potential, fatal consequences, and the inequitable distribution of risks and benefits. Factor 2, labeled "Unknown Risk," is defined

---

[1] Lincoln Moses, a statistician at Stanford University, has pointed out that this factor space summarizes the relationships among more than 1400 mean scores built from more than 40,000 judgments.

TABLE 1—*Ordering of Perceived Risk for 30 Activities and Technologies. The ordering is based on the geometric mean risk ratings within each group. Rank 1 represents the most risky activity or technology.*

	League of Women Voters	College students	Active Club members	Experts
Nuclear power	1	1	8	20
Motor vehicles	2	5	3	1
Handguns	3	2	1	4
Smoking	4	3	4	2
Motorcycles	5	6	2	6
Alcoholic beverages	6	7	5	3
General (private) aviation	7	15	11	12
Police work	8	8	7	17
Pesticides	9	4	15	8
Surgery	10	11	9	5
Fire fighting	11	10	6	18
Large construction	12	14	13	13
Hunting	13	18	10	23
Spray cans	14	13	23	26
Mountain climbing	15	22	12	29
Bicycles	16	24	14	15
Commercial aviation	17	16	18	16
Electric power (non-nuclear)	18	19	19	9
Swimming	19	30	17	10
Contraceptives	20	9	22	11
Skiing	21	25	16	30
X rays	22	17	24	7
High school & college football	23	26	21	27
Railroads	24	23	20	19
Food preservatives	25	12	28	14
Food coloring	26	20	30	21
Power mowers	27	28	25	28
Prescription antibiotics	28	21	26	24
Home appliances	29	27	27	22
Vaccinations	30	29	29	25

at its high end by hazards judged to be unobservable, unknown, new, and delayed in their manifestation of harm.

Research has shown that laypeople's risk perceptions and attitudes are closely related to the position of a hazard within the factor space. Most important is the factor Dread Risk. The higher a hazard's score on this factor (*i.e.*, the further to the right it appears in the space), the higher its perceived risk, the more people want to see its current risks reduced, and the more they want to see strict regulation employed to achieve the desired reduction in risk. In contrast, experts' perceptions of risk are not closely related to any of the various risk characteristics or factors derived from these characteristics. Instead, as noted earlier,

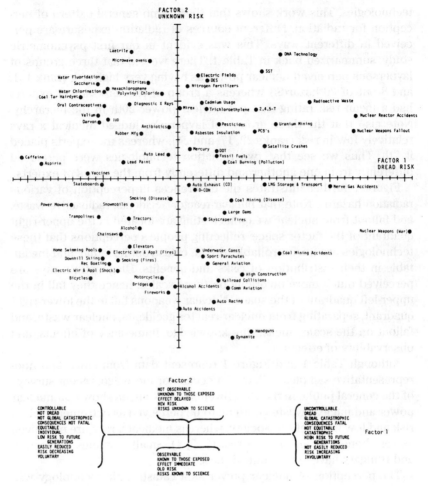

Fig. 1. Location of 31 hazards on factors 1 and 2 derived from the interrelationships among 18 risk characteristics. Each factor is made up of a combination of characteristics, as indicated by the lower diagram.

experts appear to see riskiness as synonymous with expected annual mortality. As a result, some conflicts over "risk" may result from experts and laypeople having different definitions of the concept.

Perception of Radiation Risk

Numerous psychometric surveys conducted during the past decade have examined perceptions of risk and benefit from various radiation technologies. This work shows that there is no general pattern of perception for radiation. Different sources of radiation exposure are perceived in different ways. This was evident in our first psychometric study, summarized back in Table 1. There we see that three groups of laypersons perceived nuclear power as having very high risk (rank 1, 1, and 8 out of 30 hazards) whereas a group of risk-assessment experts had a mean risk rating that put nuclear power 20th in the hierarchy. Note also that the three groups of laypersons judged medical x rays relatively low in risk (ranks 22, 17, and 24), whereas the experts placed it 7th. Thus we see that two radiation technologies were perceived differently from one another and differently from the views of experts.

Figure 1 further illustrates the differences in perception of various radiation hazards. Note that nuclear-reactor accidents, radioactive waste, and fallout from nuclear weapons testing are located in the upper-right quadrant of the factor space, reflecting people's perceptions that these technologies are uncontrollable, dread, catastrophic, lethal, and inequitable in their distribution of risks and benefits. Diagnostic x rays are perceived much more favorably on these scales, hence they fall in the upper-left quadrant of the space. Nuclear weapons fall in the lower-right quadrant, separating from nuclear-reactor accidents, nuclear waste, and fallout on the scales measuring knowledge, immediacy of effects, and observability of effects.

Although Table 1 and Figure 1 represent data from small and nonrepresentative samples collected a decade or more ago, recent surveys of the general public in the U.S., Sweden, and Canada show that nuclear power and nuclear waste continue to be perceived as extremely high in risk and low in benefit to society, whereas medical x rays are perceived as very beneficial and low in risk (2, 3, 4). Smaller studies in Norway and Hungary have also obtained these results (5).

The perception of nuclear power as a catastrophic technology was studied in depth by Slovic, Lichtenstein, and Fischhoff (6). They found that, before the TMI accident, people expected nuclear-power accidents to lead to disasters of immense proportions. Imagined scenarios of reactor accidents were found to resemble scenarios of the aftermath of nuclear war. Replication of these studies after the TMI event found even more extreme "images of disaster." The powerful negative imagery evoked by nuclear power and radiation is discussed from a historical perspective by Weart (7).

Further indication of the diversity of radiation perceptions is provided by large-scale surveys of perceptions of risk from radon in the home. Sandman, Weinstein, and Klotz (8) surveyed residents in the Reading Prong area of New Jersey, a region characterized by very high radon

TABLE 2—*Summary of perception and acceptance of risks from diverse sources of radiation exposure.*

	PERCEIVED RISK	
	Technical Experts	Public
Nuclear Power/Nuclear Waste	Moderate Risk	Extreme Risk
	Acceptable	Unacceptable
X rays	Low/Moderate Risk	Very Low Risk
	Acceptable	Acceptable
Radon	Moderate Risk	Very Low Risk
	Needs Action	Apathy
Nuclear Weapons	Moderate to Extreme Risk	Extreme Risk
	Tolerance	Tolerance
Food Irradiation	Low Risk	High Risk?
	Acceptable	Acceptability Questioned
Electric and Magnetic Fields	Low Risk	Not Yet Aware
	Acceptable	

levels in many homes. They found that residents there were basically apathetic about the risk. Few had bothered to monitor their homes for radon. Most believed that, although radon might be a problem for their neighbors, their own homes did not have any problem.

A striking contrast to the apathy regarding radon in homes is the strong public reaction that developed in many New Jersey cities when the state attempted to develop a landfill in which to place 14,000 barrels of mildly radioactive soil. The soil had been excavated from the former site of a radium watch-dial factory that had operated at the turn of the century. Over a period of several years, the state tried in vain to find a community that would accept the soil (9).

Table 2 summarizes the status of perceived risk for six radiation technologies, contrasting the views of technical experts with the views of the general public. In addition to nuclear power, nuclear waste, x rays, radon, and nuclear weapons, I have added food irradiation and a source of non-ionizing radiation, electric and magnetic fields, to the table, although there is only sparse information about perceptions of these two sources. We see that, except for nuclear weapons, there is little agreement between the experts and the public regarding the level of risk or its acceptability. To my knowledge there has been only one study thus far of perceptions of risk from electric and magnetic fields. This study, by Morgan et al. (10), found that perceived risks associated with fields from power lines and electric blankets were relatively low.

However, when the respondents were given a briefing about research on health effects of electric fields (which said that many studies had been done but no adverse human health effects had yet been reliably demonstrated), their perceptions on subsequent retest shifted towards greater perceived risk. They also saw risks from transmission lines and electric blankets as less well known, more dread, more likely to be fatal, less equitable, and less adequately controlled after receiving this information. These results imply that, if concerns about electric and magnetic fields become publicized, public fears will increase.

It is instructive to compare perceptions of risk and benefit for various radiation technologies with perceptions of various chemical technologies. Concerns about chemical risks have risen dramatically in the past decade, spurred by well-publicized crises at Love Canal, Times Beach, Missouri, and many other waste sites; by major accidents at Seveso, Bhopal, and Valdez; and by numerous other problems such as the contamination of ground water and flour with the pesticide ethylene dibromide (EDB) and the recent flap regarding the use of Alar, a growth regulator, in apples. The image of chemical technologies is so negative that when you ask members of the general public to tell you what first comes to mind when they hear the word "chemicals," by far the most frequent response is "dangerous" or some synonym (*e.g.*, toxic, hazardous, poison, deadly). Chemicals in general and agricultural and industrial chemicals in particular are seen as very high risk and very low benefit, as are nuclear power and nuclear-waste technologies. However, just as medical uses of radiation (such as x rays) are perceived in a very favorable way, differently from other radiation technologies, so are prescription drugs, which are a very potent and toxic category of chemicals to which we are often exposed at high doses. Figure 2, taken from a study in Canada (4) illustrates the parallels between nuclear power and non-medical chemicals (pesticides, food additives) seen as high in risk and low in benefit and x rays and prescription drugs (high benefit/low to moderate risk). A national survey in Sweden has shown much the same results (3).

Lessons

What does this research tell us about the acceptance of risk from radiation? There seem to be several lessons:

First, the acceptance afforded x rays and prescription drugs clearly shows that acceptance of risk is conditioned by perceptions of direct benefits and by trust in the managers of the technology, in this case the medical and pharmaceutical professions. The managers of nuclear power

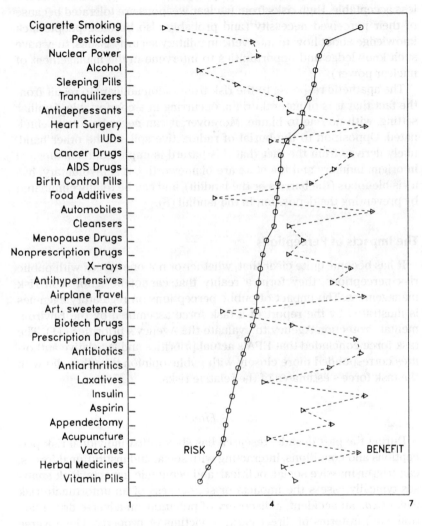

Fig. 2. Mean perceived risk and perceived benefit for 33 activities, substances, and technologies based on a national survey in Canada (4).

and other chemical technologies are clearly less trusted and the benefits of these technologies are not highly appreciated, hence their risks are less acceptable. High risks from nuclear weapons are tolerated because of their perceived necessity (and probably also because people lack knowledge about how to intervene in military security issues; they have such knowledge and opportunities to intervene in the management of nuclear power).

The apathetic response to the risk from radon appears to result from the fact that it is of natural origin, occurring in a comfortable, familiar setting, with no one to blame. Moreover, it can never be totally eliminated. Opposition to the burial of radioactive soil, on the other hand, likely derives from the fact that this hazard is imported, technological in origin, industry and the state are blameworthy, it is involuntary, has a visible focus (the barrels or the landfill), and can be totally eliminated by preventing the deposition in the landfill (8).

The Impacts of Perceptions

It has become quite clear that, whether or not one agrees with public risk perceptions, they form a reality that cannot be ignored in risk management. The impact of public perceptions on regulatory agencies is illustrated by the report of a task force assembled by the Environmental Protection Agency to evaluate the Agency's priorities (11). The task force concluded that EPA's actual priorities and legislative authorities corresponded more closely with public opinion than they did with the task force's estimates of the relative risks.

Ripple Effects

During the past decade, research has shown that individual risk perceptions and cognitions, interacting with social and institutional forces, can trigger massive social, political, and economic impacts. Risk analyses typically assess the impacts or seriousness of an unfortunate risk event (*e.g.*, an accident, a discovery of pollution, an adverse drug reaction, etc.) in terms of direct harm to victims or property. The adverse impacts of a risk event sometimes extend far beyond these direct harmful effects, and may include indirect costs to the responsible government agency or private company that far exceed direct costs. In some cases, all companies in an industry are affected, regardless of which company was responsible for the mishap. In extreme cases, the indirect costs of a mishap may even extend past industry boundaries, affecting companies, industries, and agencies whose business is minimally related to the initial event. Thus, an unfortunate event can be thought of as a stone dropped in a pond. The ripples spread outward, encompassing first the directly affected victims, then the responsible company or agency, and, in the extreme, reaching other companies, agencies, and industries.

Some events make only small ripples; others make big ones. The challenge is to discover characteristics associated with an event and the way that it is managed that can predict the breadth and seriousness of these impacts (see Figure 3). Early theories equated the magnitude of impact to the number of people killed or injured, or to the amount of property damaged. Unfortunately, things aren't this simple. The accident at the Three Mile Island (TMI) nuclear reactor in 1979 provides a dramatic demonstration that factors besides injury, death, and property damage impose serious costs. Despite the fact that not a single person died at TMI, and few if any latent cancer fatalities are expected, no other accident in our history has produced such costly societal impacts. The accident at TMI devastated the utility that owned and operated the plant. It also imposed enormous costs on the nuclear industry and on society, through stricter regulation, reduced operation of reactors worldwide, greater public opposition to nuclear power, reliance on more expensive energy sources, and increased costs of reactor construction and operation. It may even have led to a more hostile view of other large-scale, modern technologies, such as chemical manufacturing and genetic engineering. The point is that traditional economic and risk analyses tend to neglect these higher-order impacts, hence they greatly underestimate the costs associated with certain kinds of mishaps.

Although the TMI accident is extreme, it is by no means unique. Other recent events resulting in enormous higher-order impacts include the chemical manufacturing accident at Bhopal, India; the pollution of Love Canal, New York, and Times Beach, Missouri; the disastrous launch of the space shuttle Challenger; and the meltdown of the nuclear reactor at Chernobyl.

Accidents as Signals

A theory aimed at describing how psychological, social, cultural, and political factors interact to "amplify risk" and produce ripple effects has been presented by Kasperson, Renn, Slovic *et al.* (12). An important element of this theory is the assumption that the perceived seriousness of an accident or other unfortunate event, the media coverage it gets, and the long-range costs and other higher-order impacts on the responsible company, industry, or agency are determined, in part, by what that event signals or portends. *Signal value* reflects the perception that the

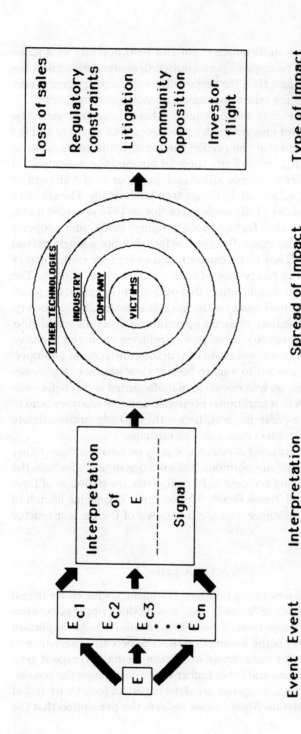

Fig. 3. A preliminary model of impact for unfortunate events. Development of the model will require knowledge of how the characteristics (E_c) associated with a hazard event interact to determine the interpretation or message drawn from that event. The nature of the interpretation is presumed to determine the type and magnitude of ripple effects.

ACCIDENTS AS SIGNALS

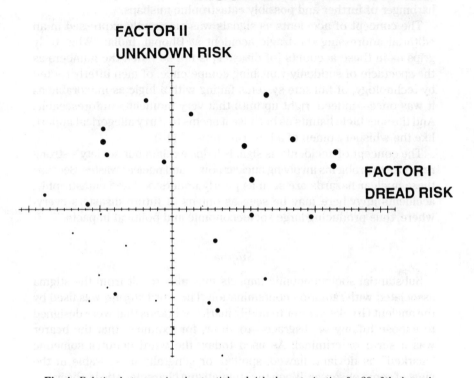

Fig. 4. Relation between signal potential and risk characterization for 30 of the hazards shown in Figure 1. The larger the point, the greater the degree to which an accident involving that hazard was judged by raters to "serve as a warning signal for society, providing new information about the probability that similar or even more destructive mishaps might occur within this type of activity." The ripple effects associated with a mishap are likely to be correlated with signal potential.

event provides new information about the likelihood of similar or more destructive future mishaps.

The informativeness or signal value of an event, and thus its potential social impact, appears to be systematically related to the characteristics of the hazard and the location of the event within the factor space described earlier (see Figure 4). An accident that takes many lives may produce relatively little social disturbance (beyond that caused the victims' families and friends) if it occurs as part of a familiar and well-understood system (*e.g.*, a train wreck). However, a small accident in an unfamiliar system (or one perceived as poorly understood), such as a nuclear reactor, a recombinant DNA laboratory, or even a prescription drug, may have immense social consequences if it is perceived as a harbinger of further and possibly catastrophic mishaps.

The concept of accidents as signals was eloquently expressed in an editorial addressing the tragic accident at Bhopal, India: "What truly grips us in these accounts [of disaster] is not so much the numbers as the spectacle of suddenly vanishing competence, of men utterly routed by technology, of fail-safe systems failing with a logic as inexorable as it was once—indeed, right up until that very moment—unforeseeable. And the spectacle haunts us because it seems to carry allegorical import, like the whispery omen of a hovering future" (13).

The concept of accidents as signals helps explain our society's strong response to problems involving nuclear power and nuclear wastes. Because these nuclear hazards are seen as poorly understood and catastrophic, accidents anywhere may be seen as omens of future disasters everywhere, thus producing large socioeconomic and political impacts.

Stigma

Substantial socioeconomic impacts may also result from the stigma associated with radiation contamination. The word *stigma* was used by the ancient Greeks to refer to bodily marks or brands that were designed to expose infamy or disgrace—to show, for example, that the bearer was a slave or criminal. As used today, the word denotes someone "marked" as deviant, flawed, spoiled, or generally undesirable in the view of some observer. When the stigmatizing characteristic is observed, the person is denigrated or avoided. Prime targets for stigmatization are members of minority groups, the aged, homosexuals, drug addicts, alcoholics, and persons afflicted with physical or mental disabilities and deformities.

A dramatic example of stigmatization involving radiation occurred in September, 1987, in Goiania, Brazil, where two men searching for scrap metal dismantled a cancer therapy device in an abandoned clinic. In doing so, they sawed open a capsule containing 28 grams of cesium chloride. Children and workers nearby were attracted to the glowing material and began playing with it. Before the danger was realized, several hundred people became contaminated and four persons eventually died from acute radiation poisoning. Publicity about the incident led to stigmatization of the region and its residents (14). Hotels in other parts of the country refused to allow Goiania residents to register, airline pilots refused to fly with Goiania residents on board, automobiles driven by Goianians were stoned, hotel occupancy in the region dropped 60%

for six weeks following the incident, and virtually all conventions were canceled during this period. The sale prices of products manufactured in Goiania dropped by 40% after the first news reports and remained depressed for a period of 30–45 days despite the fact that no items were ever found to have been contaminated.

Stigma and nuclear waste. Stigmatization also appears relevant to concerns about the potential for the high-level nuclear waste repository proposed for siting at Yucca Mountain, Nevada to have adverse economic effects on the city of Las Vegas and the Southern Nevada region. Specific impacts of concern include reduction in short-term visits to the city and region by vacationers and conventioners, effects on long-term residents (emigration, reduced immigration of retirees, and reduced ability to attract new businesses).

Assessment of these impacts is obviously important to citizens and officials of Nevada, who need to know what adverse economic consequences to expect if Yucca Mountain is developed as the repository. Indeed, selection of Yucca Mountain as the prime candidate and attempts to evaluate its qualifications over the next few years may trigger some of these impacts in advance of the final decision. Information about possible economic impacts may be relevant to the final decision itself. Moreover, such information is essential for decisions about compensation and mitigation.

My colleagues and I have addressed this question through the study of stigma and imagery (15). By means of an extensive series of empirical studies we showed that:

1. People have consistent, often stereotypical, images of environments (*e.g.*, Los Angeles is perceived as smoggy, sunny, and crowded; Las Vegas evokes images of gambling, entertainment, bright lights, and money).
2. These images have diverse positive and negative affective meanings that influence preferences for environments (*e.g.*, in this case, preferences for sites in which to vacation, retire, find a job, or start a new business). In other words, imagery is highly predictive of preferences and behaviors.
3. The concept of a nuclear-waste repository evokes a wide variety of strongly negative images (see Table 3), consistent with extreme perceptions of risk and stigmatization.
4. The nuclear weapons test site near Yucca Mountain, which has been around far longer than the nuclear waste project, has led to a modest amount of nuclear imagery being associated with the State of Nevada. Persons who exhibited such nuclear images

TABLE 3—*Hierarchy of Images Associated with an "Underground Nuclear Waste Storage Facility".*

Category	Frequency	Images Included in Category
1. Dangerous	179	dangerous, danger, hazardous, toxic, unsafe, harmful, disaster
2. Death	107	death, sickness, dying, destruction
3. Negative	99	negative, wrong, bad, unpleasant, terrible, gross, undesirable, awful, dislike, ugly, horrible
4. Pollution	97	pollution, contamination, leakage, spills, Love Canal
5. War	62	war, bombs, nuclear war, holocaust
6. Radiation	59	radiation, nuclear, radioactive, glowing
7. Scary	55	scary, frightening, concern, worried, fear, horror
8. Somewhere Else	49	wouldn't want to live near one, not where I live, far away as possible
9. Unnecessary	44	unnecessary, bad idea, waste of land
10. Problems	39	problems, trouble
11. Desert	37	desert, barren, desolate
12. Non-Nevada Locations	35	Utah, Arizona, Denver
13. Storage Location	32	caverns, underground salt mines
14. Government/Industry	23	government, politics, big business

were much less likely than others to prefer Nevada over competing sites.

The repository is a high-signal hazard. It is likely to receive extensive media coverage, even in the event of minor mishaps or controversies. Over time, this could lead to association of the repository with the region of Southern Nevada—an association that is likely to affect people's behaviors in ways that inflict economic damages on the region. Predicting the precise magnitude and duration of these impacts is, however, impossible. The uncertainties involved in repository development and operation make it inevitable that the actual impacts—physical, biological, social, and economic—will differ from the best of impact projections.

The disposal of nuclear wastes is a technology that the nuclear industry believes can be managed safely and effectively. The discrepancy between this view and the images shown in Table 4 is indeed startling.

Risk Communication: Placing Radiation Risks in Perspective

Given the importance of risk perceptions and the extraordinary divergence between perceptions of experts and laypersons in the domains of chemical and radiation technologies, it is not surprising that there has been a burgeoning interest in the topic of "risk communication." Much has been written about the need to inform and educate people about risk and the difficulties of doing so (16–24). As many writers have observed, doing an adequate job of communicating about risk means finding comprehensible ways of presenting complex technical material that is clouded by uncertainty and is inherently difficult to understand.

The crux of the communication problem is providing information that puts risk into perspective in a way that facilitates decision making. One important lesson emerged from the controversy over ethylene dibromide, EDB. The Environmental Protection Agency, which was responsible for regulating EDB, disseminated information about the aggregate risk of this pesticide to the exposed population. Although the media accurately transmitted EPA's "macro" analysis, newspaper editorials and public reaction clearly indicated an inability to translate this into a "micro" perspective on the risk to an exposed individual. What the newspaper reader or TV viewer wanted to know, and had trouble learning, was the answer to the question "Should I eat the bread?" (25).

One of the few "principles" in this field that seems to be useful is the assertion that comparisons are more meaningful than absolute numbers or probabilities, especially when these absolute values are quite small. Sowby (26) argued that to decide whether or not we are responding adequately to radiation risks we need to compare them to "some of the other risks of life." Rothschild (27) observed "There is no point in getting into a panic about the risks of life until you have compared the risks which worry you with those that don't, but perhaps should."

Typically, such exhortations are followed by elaborate tables and even "catalogs of risks" in which diverse indices of death or disability are displayed for a broad spectrum of life's hazards. Thus Sowby (26) provided extensive data on risks per hour of exposure, showing, for example, that an hour riding a motorcycle is as risky as an hour of being 75 years old. Wilson (28) developed a table of activities (*e.g.*, flying 1000 miles by jet, having one chest x ray), each of which is estimated to increase one's annual chance of death by 1 in one million (which in the case of accidental death would decrease one's life expectancy by about 15 minutes). In similar fashion, Cohen and Lee (29) ordered many hazards in terms of their reduction in life expectancy on the assumption that "to some approximation, the ordering should be society's order of priorities. However, we see several very major problems that have received very little attention ... whereas some of the items near the bottom of the list, especially those involving radiation, receive a great deal of attention" (29, p. 720). A related exercise by Reissland and Harries (30) compared loss of life expectancy in the nuclear industry with that in other occupations.

Although such risk comparisons may provide some aid to intuition, they do not educate as effectively as their proponents have assumed. For example, though some people may feel enlightened upon learning that a single takeoff or landing in a commercial airliner takes an average of 15 minutes off one's life expectancy, others may find themselves completely bewildered by such information. When landing or taking off, one will either die prematurely (almost certainly by more than 15 minutes) or one will not. From the standpoint of the individual, averages do not adequately capture the essence of such risks.

Furthermore, the research on risk perception described earlier shows that perception and acceptance of risk are determined not only by accident probabilities, annual mortality rates, and losses of life expectancy, but also by numerous other characteristics of hazards such as uncertainty, controllability, catastrophic potential, equity, and threat to future generations. Within the perceptual space defined by such characteristics, each hazard is unique. A statement such as "the annual risk from living near a nuclear power plant is equivalent to the risk of riding an extra 3 miles in an automobile" fails to consider how these two technologies differ on many qualities that people believe to be important. As a result, such statements are likely to produce anger rather than enlightenment.

In sum, comparisons across diverse hazards may be useful tools for educating the public. Yet the facts do not speak for themselves. Comparative analyses must be performed with great care to be worthwhile.

Fortunately, radiation risks are relatively well understood. Radiation emissions can be measured precisely and comparisons can be made between actual or potential exposure levels of concern and familiar, everyday exposures from natural sources of radiation or medical x rays and treatments. By making comparisons from one source of radiation to another, one avoids the apples vs. oranges comparisons that befuddle and anger people.

Wilson (31) used comparisons with natural sources of radiation to put the risks from the Chernobyl accident into perspective for the 2 million people living downwind from the reactor in Byelorussia. He noted that the estimated increased lifetime dose was 0.7 rem for each of these persons and that this is considerably less than the difference in the lifetime external dose a person receives on moving from New York to Denver. It is also less than the difference in the dose a person receives from inhaled radon if he or she moves from an average New England house to an average Pennsylvania house.

When radiation from Chernobyl reached the United States, the Inter-Agency Task Force, chaired by EPA administrator Lee Thomas, used similar comparisons to illustrate the low level of risk involved. Media stories pointed out that exposures in the U.S. were a small fraction of the exposure from a chest x ray. A news story from Portland, Oregon indicated that readings of 2.9 picocuries of iodine-131 per cubic meter of air were insignificant compared to the 2700 picocurie level that would trigger concern.

This discussion is not meant to imply that we already know how to communicate radiation risks effectively. Communication about Chernobyl was dreadful in Europe (32, 33). Information messages were prepared with different terms (roentgens, curies, bequerels, rads, rems, sieverts, grays) which were explained poorly or not at all. Public anxiety was high and not related to actual threat. Public officials were at odds with one another and inconsistent in their evaluations of risks from consuming various kinds of food or milk. Comparisons with exposure to natural radiation from familiar activities were not well received because the media and the public did not trust the sources of such information. Other comparisons (*e.g.*, with background cancer rates) fared even worse. Many of the statements made by officials to calm the public confused and angered them instead. Although communications in the U.S. effectively maintained a calm perspective, one could say that U.S. officials had a relatively easy job. All they had to do was convince people that minuscule levels of radiation were not a threat. Had there been higher levels and "hot spots" as in the Soviet Union and Western Europe, the job of communicating would have been far tougher and it is not clear that proper perspectives on risk would have been achieved.

The good news is that enough is known about radiation and about risk communication to enable us to craft good risk comparisons, if we devote proper attention and resources to doing so (see, *e.g.*, the effort to inform homeowners about their risks from radon (34); and the recommendations by Adelstein (35)). Radiation health experts, emergency-response specialists, and risk-communication experts should be brought together in order to produce a guide to risk communication in the event of a major radiation release in the U.S. This guide should be produced *now*, before the next release occurs. Without it, effective communication about risk is unlikely.

Concluding Comments

Ten years post TMI:

1. Nuclear power is still perceived as an extraordinarily risky technology. There is still no solution in the U.S. to the problem of

disposing of nuclear wastes—the siting of a repository for permanent disposal in Nevada is being actively opposed by the public and the government in that state and by interest groups elsewhere. However, there is also rapidly escalating awareness and concern of the environmental and health risks of burning fossil fuels. Obviously we will face some difficult choices in the years ahead.
2. Nuclear power is no longer uniquely feared. During the past decade many chemical technologies have become similarly stigmatized. What technologies are next in line? Perhaps they will be certain applications of biotechnology.
3. Despite billions of dollars expended each year in managing risks, citizens of the United States, Canada, and many other countries see themselves more at risk from technology than ever before. The benefits of technologies (except for medical uses) are taken for granted. Zero risk, at any cost, is a desired goal for many.
4. Quantitative risk assessment within the nuclear and chemical industries has had little impact on public attitudes except, perhaps to raise concerns.
5. There has been little serious research on how to improve communication about radiation risks, except in the area of radon.

Have we progressed?

I believe that conflicts involving the perception and acceptance of risk from radiation, chemical, and biological technologies represent a major societal crisis. Multidisciplinary, multi-industry, and multi-partisan groups need to join forces to face this problem in a constructive way. Perhaps if we take this task seriously, we will make greater progress during the next decade.

References

1. Slovic, P. "Perception of risk," Science **236,** 280–285 (1987).
2. Kunreuther, H., Devousges, W.H. and Slovic, P. "Nevada's predicament: Public perceptions of risk from the proposed nuclear waste repository," Environment **30**(8), 16–20, 30–33 (1988).
3. Slovic, P., Kraus, N., Lappe, H., Letzel, H. and Malmfors, T. "Risk perception of prescription drugs: Report on a survey in Sweden," Pharmaceutical Medicine **4,** 43–65 (1989).
4. Slovic, P., Kraus, N., Lappe, H. and Major, M. *Risk perception of prescription drugs: Report on a survey in Canada* (Report No. 89-4) (Decision Research, Eugene, Oregon) (1989).

5. Teigen, K.H., Brun, W. and Slovic, P. "Societal risks as seen by a Norwegian public," Journal of Behavioral Decision Making **1**, 111–130 (1988).
6. Slovic, P., Lichtenstein, S. and Fischhoff, B. "Images of disaster: perception and acceptance of risks from nuclear power," page 223–245 in *Energy Risk Management* Goodman, G. and Rowe, W. Eds. (Academic Press, London) (1979).
7. Weart, S. *Nuclear fear: A history of images* (Harvard University, Cambridge, Massachusetts) (1988).
8. Sandman, P.M., Weinstein, N.D. and Klotz, M.L. "Public response to the risk from geological radon," Journal of Communication **37**, 93–108 (1987).
9. Carlson, E. "Suburban radium lode gives New Jersey a headache," Wall Street Journal page 23 (December 23, 1986).
10. Morgan, M.G., Slovic, P., Nair, Il, Geisler, D., MacGregor, D., Fischhoff, B., Lincoln, D. and Florig, K. "Powerline frequency electric and magnetic fields: A pilot study of risk perception," Risk Analysis **5**, 139–149 (1985).
11. Allen, F.W., "The situation: What the public believes: How the experts see it," EPA Journal, **13**(9), 9–12 (1987).
12. Kasperson, R.E., Renn, O., Slovic, P., Brown, H.S., Emel, J., Gobel, R., Kasperson, J.X. and Ratick, S. "The social amplification of risk: A conceptual framework." Risk Analysis **8**, 177–187 (1988).
13. Talk of the town. New Yorker, February 18, 1985.
14. Petterson, J.S. "Perception vs. reality of radiological impact: The Goiania model," Nuclear News **31**(14), 84–90 (1988).
15. Slovic, P., Layman, M., Kraus, N., Chalmers, J., Gesell, G. and Flynn, J. "Perceived risk, stigman, and potential economic impacts of a high-level nuclear waste repository in Nevada (Report No. 89-3) (Decision Research, Eugene, Oregon) (1989).
16. Covello, V.T., Sandman, P.M. and Slovic, P. "Risk communication, risk statistics, and risk comparisons: A manual for plant managers," (Chemical Manufacturers Association, Washington)
17. Covello, V.T., von Winterfeldt, D. and Slovic, P. "Risk communication: A review of the literature," Risk Abstracts **3**, 171–182 (1986).
18. Covello, V.T., von Winterfeldt, D. and Slovic, P. "Risk communication: research and practice," Unpublished manuscript (Columbia University, School of Public Health, New York) (1988)
19. Hance, B.J., Chess, C. and Sandman, P.M. "Improving dialogue with communities: A risk communication manual for government risk communication," (Department of Environmental Protection, Division of Science and Research, Trenton, New Jersey) (1988).
20. Krimsky, S. and Plough, A. "Environmental Hazards: Communicating risks as a social process," (Auburn House, Dover, Massachusetts) (1988).
21. Sandman, P.M. "Explaining environmental risk," Report No. TS-799 (Environmental Protection Agency, Office of Toxic Substances, Washington) (1986).
22. Slovic, P. "Informing and educating the public about risk," Risk Analysis **4**, 403–415.

23. Slovic, P., Fischhoff, B. and Lichtenstein, S. "Informing people about risk," page 165–181 in *Product Labeling and Health Risks* (Report No. 6) Morris, L., Mazis, M. and Barofsky, I., Eds. (The Banbury Center, Cold Spring Harbor, New York) (1980).
24. Slovic, P., Fischhoff, B. and Lichtenstein, S. "Informing the public about risks of ionizing radiation," Health Phys. **41**, 589–598 (1981).
25. Sharlin, H.I. "EDB: A case study in the communication of health risk," Risk Analysis **6**, 61–68 (1986).
26. Sowby, F.D. "Radiation and other risks," Health Phys. **11**, 879–887 (1965).
27. Rothschild, N. "Coming to grips with risk," (Address presented on BBC television: reprinted in the Wall Street Journal) (November, 1978).
28. Wilson, R. "Analyzing the daily risks of life," Technological Review **81**, 40–46 (1979).
29. Cohen, B. and Lee, I. "A catalog of risks," Health Phys. **36**, 707–722 (1979).
30. Reissland, J., and Harries, V. "A scale for measuring risks," New Scientist **83**, 809–811 (1979).
31. Wilson, R. Testimony before the Subcommittee on Nuclear Regulation, Committee on the Environment and Public Works. U.S. Senate, Washington, D.C. (May 7, 1987).
32. Otway, H., Haastrup, P., Connell, W., Gianitsopoulas, G. and Paruccini, M. "Risk communication in Europe after Chernobyl: A media analysis of seven countries," Industrial Crisis Quarterly **2**, 31–35 (1988).
33. Wynne, B. "Sheep farming after Chernobyl: A case study in communicating scientific information," Environment **31**(2), 10–15; 33–39 (1989).
34. Johnson, F.R., Fisher, A., Smith, V.K. and Desvousges, W.H. "Informed choice or regulated risk? Lessons from a study in radon communication," Environment **30**(4), 12–15; 30–35 (1988).
35. Adelstein, S.J. "Uncertainty and relative risks of radiation exposure," J. of the Amer. Med. Assoc. **258**, 655–657 (1987).

Discussion

MERRIL EISENBUD: I like, Paul, what you've described as the bottom line as where we stand, namely how do people think. And it's time we gave more consideration to why it is that they think that way. I think most of us know. There are some things that are commonly accepted and you sort of went along with it with the implication that there are more disasters now. There aren't more disasters, if you look at the table of accidents in the world almanac and list the number of mine cave-ins that killed over 100 people in the first fifty years of this century and second, it's something like 15 to 1 ship sinkings. Chemical explosions were all far worse. We learned to live with those. Now the only thing that has happened is that we have a new modality that's about forty years old, namely television. That is able to condition people to think thoughts about which automobile to buy, which depilatory to use, which pantyhose to put on, which corn flakes to buy, and it's teaching them how to think about these risks that you're talking about.

PAUL SLOVIC: Well, obviously the media is critical. We get most of our information about technological risks from the media. They play a role, but it's not the only factor. I think we have to look at a lot of fundamental aspects of our society. The fact that our society is adversarial in its approach to risk management. And we fight things out in the courts. When we litigate, we have experts on both sides of the issue contradicting each other and making expertise and risk analysis look foolish. And all of this is highly publicized by the media and makes the public wonder if anyone knows what's going on. So it's a combination of media and other things that are fundamental to this society. I think we know how people think about risk, I mean they respond to experience and to the ease with which they can imagine things happening. They don't necessarily think like scientists. Even though there may be fewer disasters now than in the past, whatever there is, is highly publicized. The successes of technologies are less well publicized. It seems only where we have a personal relationship and a high perceived benefit such as in the medical arena, that we have a different conception of the radiation and chemical technologies. So perhaps that provides the clue to dealing with these technologies. I think that we have to impress upon people the benefits, if indeed there are benefits, and then hope that the people can make wise decisions, because our society is one in which the people do have a great say in these decisions.

ROBERT BRENT (Jefferson Medical College, Philadelphia): I don't want to complicate such a complicated subject, but I'm just wondering whether you would briefly comment on one other factor, and that is that the population you're sampling is not homogeneous and one of the factors is that in a population like in our country, significant numbers of them do have emotional problems. Anxiety, probably an emotional illness, is a very serious problem in most civilized cultures and the impact of anxiety, its treatment, and the response to these signals and the possibility that a small percentage could amplify or have impact on the rest of the population is very difficult to study. But I think its another area of research because you can even see a scientist who has an emotional immaturity or anxiety have a tremendous impact when his own inability to handle personal problems gets involved in his scientific work and he can have an impact, because of his credentials, on a significant part of the population.

PAUL SLOVIC: Well, I think that technology does make people anxious but I don't see anxiety as the key factor here. I think that what we see are people responding to clues in their environment. They're responding in a very natural way given where they're coming from. And when you read in the paper about two experts contradicting each other over a technology it's not irrational or emotional to think that maybe expertise isn't on top of this. Or when you read about industries which have polluted in various ways and are definitely at fault, it's not irrational or emotional to be suspect of industry and its willingness to put public health ahead of its profits. Again, these are the kinds of experiences that we get through the media. And that's what people are responding to. So while they may become anxious, I don't think that anxiousness or emotionality is the problem. This is just the way people are intuitive toxicologists, intuitive radiation health people; and they use the cues that they get from their environment. And they're using them in a way that's understandable even if you don't agree with it from a scientific perspective.

NO NAME: Going to the risk communication issue, at least as I understand communication sciences, the most effective communication is when the parties are hearing each other and they understand the terms. Miller, at Northern Illinois University, has been doing a lot of work in which he has found that a very small percentage of the survey population are what he defines as scientifically literate. E.D. Hurson's well known book, "Cultural Literacy," similarly talks about the inability of a large population or portion of the population understanding very fundamental terms. I wonder whether the public really understands what we mean when we talk about risk? Do they understand what we talk about when we use the term probability? And although, we in this room, and other experts are well familiar with these terms, when we talk to the general public, they may not even know what we're saying. Is this a fundamental problem in your own mind and would an appropriate strategy be to try to convince some fourteen thousand school boards in this country to try to redefine curricula so that these very fundamental issues can be incorporated?

PAUL SLOVIC: Yes, and yes to those questions. Obviously scientific literacy is low and in a country where the public has great say they have a responsibility to at least understand as much as possible about the technologies that they're voting yes or no on. We know that the literacy is low. We know that probability and risk are very difficult concepts for people to understand. They're very difficult for scientists to understand too; if you look at the scientific literature and controversies, arguments over Bayesian versus non-Bayesian statistics and the like, you see that scientists also have difficulties struggling with concepts of risk and probability. But it certainly wouldn't hurt if we knew a little bit more about what we're talking about when we make decisions about technologies.

SCIENTIFIC SESSION

Radiation Protection Experience

Charles Meinhold
Chairman

Occupational Exposures and Practices in Nuclear Power Plants

John Baum[1]
Brookhaven National Laboratory
Upton, New York

Introduction

As the National Council on Radiation Protection and Measurements (NCRP) celebrated its 60th year of service, the Nuclear Power Industry completes its "first generation" of production. The first U.S. plant to generate significant amounts of electricity was the Shippingport Plant which was a government-owned reactor operated by Duquesne Light Company. It began operations as a pressurized water reactor (PWR) in December 1957, was converted to a light water breeder reactor in 1977, and was retired in October, 1982. Other demonstration type plants with a fairly long history of operation include the Dresden 1 plant, a 207 MWe boiling water reactor (BWR) plant operated by the Commonwealth Edison Company in Illinois from August 1960 until it was shut down in 1978; the Big Rock Point Plant in Michigan, a 69 MWe BWR which was placed in operation by Consumers Power Company in December 1965; and the Yankee Row Plant, a 175 MWe PWR that has been operated by Yankee Atomic Electric in Massachusetts since July 1961. A few other small demonstration plants were operated in the early 1960's for very short periods. As of July 1988, the United States has 110 operable nuclear power plants that can generate about 102 GWe units of electricity (1).

---

[1]This work was performed under the auspices of the U.S. Nuclear Regulatory Commission.

Of these, 36 are licensed BWR plants, one is an unlicensed BWR plant, one is a fast breeder reactor, one is a licensed high-temperature gas-cooled reactor, which has operated intermittently since July 1984 and may be shut down permanently, and 71 are licensed PWR plants. There are 9 PWR and 2 BWR plants under construction. Twenty units including those mentioned above have been shut down over the past 30 years. These include several small units used for research and development which generated 5 to 75 MWe units of power, and several larger plants such as Dresden 1, mentioned above. Others were the Hanford N reactor, a light-water graphite-moderated unit capable of generating 860 MWe units of electricity; Indian Point 1, a PWR rated at 277 MWe; and Three Mile Island 2, the PWR which was closed after the well-known March 1979 accident.

As the first generation of commercial nuclear power comes to a close, it is timely to consider the status of occupational exposure in the power generation industry, that is, the collective occupational radiation doses received by workers in nuclear power plants. The picture is surprising. One might have thought that as newer, larger, and more modern plants came on line, there would be a significant decrease in exposure per unit of electricity generated. There is some indication that this is now happening. One might also have thought that the United States, being a leader in the development of nuclear power, and in the knowledge, experience and technology of nuclear radiation protection, would have the greatest success in controlling exposure. This expectation has not been fulfilled (2).

In the early 1980's, the U.S. exposures were increasing proportionally more than increases in electricity generated. Therefore, the Nuclear Regulatory Commission began funding studies to identify the main causes for the exposures, and possible techniques for dose equivalent reduction; to compare U.S. experiences to those in other developed countries; and to provide examples of cost-benefit studies to help judge if doses were "as low as reasonably achievable" (ALARA), a basic element in the system of radiation protection. This review will rely heavily on the findings from these studies.

Figure 1 summarizes the history of the U.S. occupational collective dose equivalent for the major industries (3). Although some segments of the exposed population have shown decreased doses, the contribution from the nuclear fuel cycle has increased from an almost insignificant level in 1960 to become the dominant contributor to occupational doses in 1985. This large increase is due primarily to the large increase in number of commercial operating nuclear power plants, but also to the greater maintenance required on older plants and a number of safety

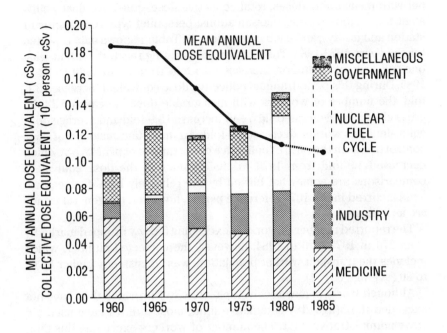

MEAN ANNUAL DOSE AND COLLECTIVE DOSE TO POTENTIALLY
EXPOSED WORKERS, 1960 TO 1985.

Fig. 1. Mean annual dose equivalent and collective dose equivalent to potentially exposed workers, 1960 to 1985.

related actions mandated by the Nuclear Regulatory Commission. Considering all industries, the mean annual dose equivalent per worker has decreased from about 0.18 cSv in 1960 to about 0.1 cSv in 1985. However, mean annual dose equivalent to nuclear power plant workers has averaged about five times these values.

Exposures in the United States

Each year the NRC requires its licensees to report their data on the occupational exposure of both employees and temporary workers. This detailed data is summarized in an annual report that provides information on the electricity generated, facility availability, number of personnel with measurable doses, total collective dose, collective dose equivalent for various worker classifications, personnel type (contractor or station and utility), and 6 job categories (4). Table 1 shows data reported for the years 1973-1985. The number of operating reactors with at least one full year of operation increased 3.4-fold from 24 in 1960 to 82 in 1985. During this period, annual collective dose equivalent increased 3.1-fold, the number of workers with measurable dose increased 6.3-fold, gross electricity generated (MWe-y) increased 5.9-fold and average dose equivalent per worker decreased 2-fold. An important measure of dose control trends in the ratio of collective dose equivalent per MWe-y, which decreased 1.9-fold from 1973 to 1985. Values in the dose equivalent comparisons are somewhat biased by the relatively high dose equivalents incurred in the 1973 reference period, however, the general trends are as indicated.

The reported number of workers exceeding 5 cSv/y dropped markedly from 270 in 1977 to 0 in 1984, however, the corrected number, which includes the transient worker population, was consistently higher by 60 to 80 (average 66) (4).

Although workers may receive up to 12 cSv/y provided they have not exceeded the 5(age -18) cumulative limit, most utilities have used a 5 cSv/y administrative limit. The number of workers exceeding this limit is not easily determined with the present reporting system since each station reports the number of workers and dose received at the station (a station may have more than one nuclear power plant on the site). Transient workers, that receive most of the collective dose, may accumulate dose at two or more sites. The number of workers receiving more than 5 cSv is thus greater than the number reported by the stations. Table 2 shows corrected numbers for number receiving > 5 cSv for the years 1977-1984 (4).

Figure 2 shows total collective dose equivalent per year, and total gross nuclear electricity generated for all U.S. nuclear power plants combined, from 1969 through 1987. As can be seen in Figure 2, as more plants were put on line in the 1970's the collective dose equivalent averaged about one person-cSv (one person-rem) per gross MWe-y generated, and the dose equivalent increased approximately in proportion to electricity generated until 1979, the year of the accident at Three Mile Island 2. In that year, and for several years thereafter, the collective dose equivalent was much more than one person-cSv per MWe-y generated, reaching a maximum of 1.85 in 1980, and a minimum of 0.79 in 1987. From 1980 through 1988, the average collective dose equivalent per operating U.S. plant declined from 1,230 cSv to 511 cSv (5).

TABLE 1—*Summary of Annual Information Reported by Commercial Light-Water-Cooled Reactors[+] 1973–1985*

Year	No. of Reactors Included	Annual Collective Doses (person-rem or person-cSv)	No. of Workers With Measurable Doses	Gross Electricity Generated (MW·y)	Average Dose Per Worker (rem or cSv)	Average Collective Dose Per Reactor (person-rem or person-cSv)	Average No. Personnel With Measurable Doses Per Reactor	Average Collective Dose per MW·y	Average Electricity Generated Per Reactor (MW·y)	Average Rated Capacity Net (MWe)
1973	24	13,963	14,780	7,164	0.94	582	616	1.9	299	496
1974	34	13,722	18,466	10,883	0.74	404	543	1.3	320	575
1975	44	20,879	25,489	17,769	0.82	475	579	1.2	404	630
1976	53	26,433	35,447	21,911	0.75	499	669	1.2	413	663
1977	57	32,511	42,266	26,444	0.77	570	742	1.2	462	677
1978	64	31,809	45,998	31,614	0.69	497	719	1.0	494	702
1979	67	39,981	64,122	29,920	0.62	597	956	1.3	447	705
1980	68	53,796	80,331	29,155	0.67	791	1,181	1.8	429	699
1981	70	54,142	82,183	31,451	0.66	773	1,174	1.7	449	719
1982	74	52,190	84,382	32,795	0.62	705	1,139	1.6	443	738
1983	75	56,471	85,646	32,926	0.66	753	1,142	1.7	439	742
1984	78*	55,214	98,092	36,441	0.56	708	1,258	1.5	467	776
1985	82**	43,042	92,871	41,601	0.46	525	1,132	1.0	507	806

[+]Includes only those reactors that had been in commercial operation for at least one full year as of December 31 of each of the indicated years, and all figures are uncorrected for multiple reporting of transient individuals.

*In 1984 it was decided that Humboldt Bay and Indian Point 1 would not be put in commercial operation again, and they are no longer included in this count of reactors.

**In 1985 it was decided that Dresden 1, a plant that has been shut down since 10/78, would not be put in commercial operation again, and it is not included in this count of reactors.

TABLE 2—*Annual Whole Body Doses Exceeding Five cSv (rem) at Nuclear Power Facilities*

Year	Reported number >5 cSv (rem)	Corrected number >5 cSv (rem)	Percent of workers
1977	270	351	0.9
1978	103	158	0.4
1979	130	180	0.3
1980	311	391	0.5
1981	189	235	0.3
1982	74	135	0.2
1983	85	168	0.2
1984	0	71	<0.1

Much of the increase in dose equivalent during the peak years following the accident at Three Mile Island 2 was attributed to improved safety actions, such as stricter environmental qualification for instrumentation, improved fire and seismic safety, and a series of "in-service inspections" on all welds in safety systems including steam pipes and pressure vessels. These in-service inspections are required once each ten years of plant operation, in addition to numerous other inspections required more frequently.

Since about 1983 the NRC and the Institute of Nuclear Power Operations (INPO, industry's self-inspection and self-regulation organization) have put increasing pressures on U.S. nuclear power plant organizations to improve training, procedures, and equipment to reduce exposures and strengthen their ALARA efforts. These efforts seem to have contributed to the reductions in exposure experienced recently.

During the period 1981-1985, the average collective dose equivalent per MWe-y for U.S. BWRs that had completed five full years of commercial operation varied from 1.5 to 7.1 person-cSv; for PWRs it varied from 0.2 to 1.0 (4). Four U.S. PWRs were below 0.3 person-cSv/MWe-y, which corresponds to the national goal in Sweden.

Comparisons with other Developed Countries

Several comparisons have been made between collective doses in U.S. plants and those in other developed countries (2,6,7). Figure 3 illustrates typical results. The collective dose equivalent for the plants in each country has been averaged over the five-year period 1978-1980 to minimize transient effects due to short-term problems with one or two plants. Thus, the data are a fairly good reflection of variations from country to country. It is remarkable that differences of more than ten-fold are

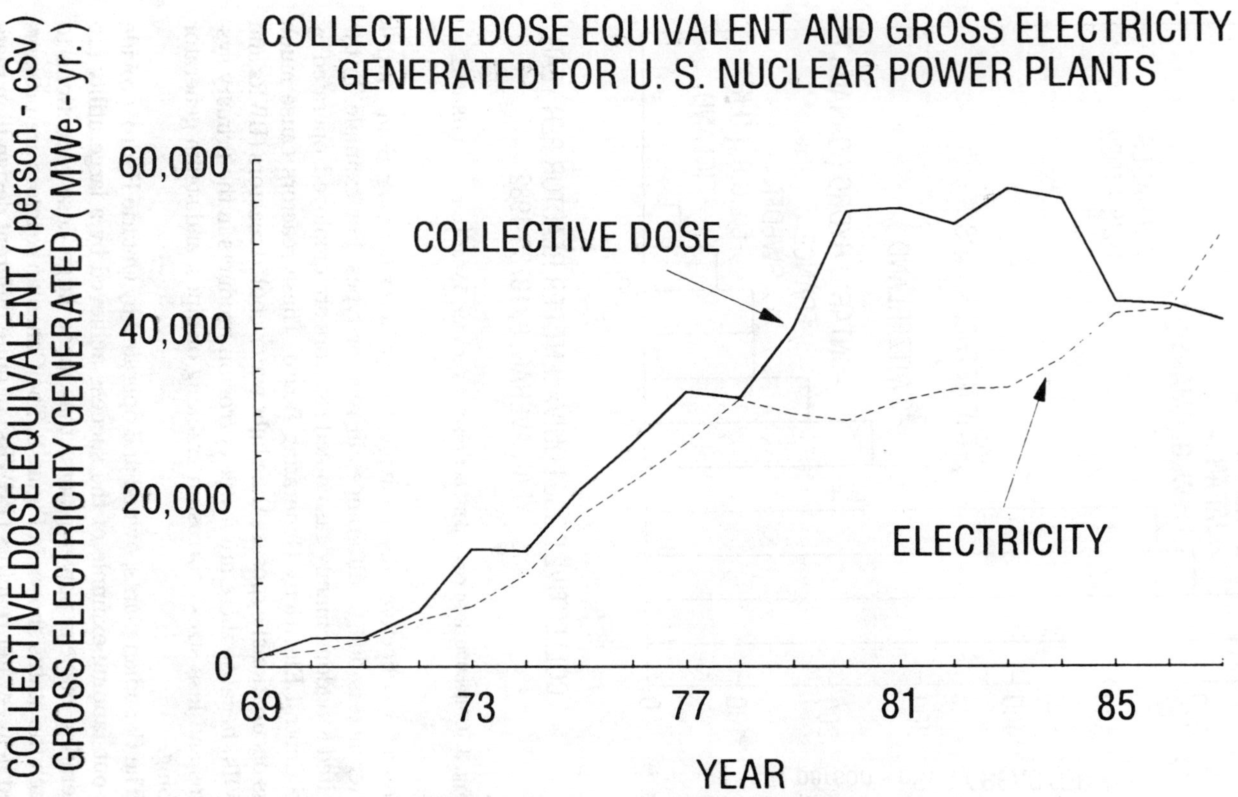

Fig. 2. Collective dose equivalent and gross electricity generated per year by U.S. nuclear power plants.

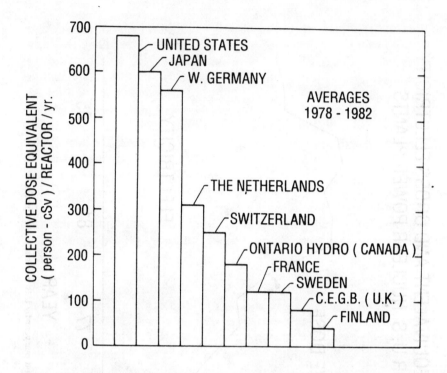

Fig. 3. Collective dose equivalent per reactor per year, 5-year averages, 1978-1982.

shown for collective dose equivalent per reactor per year. Some of this difference is due to differences in reactor types. For example, in the United Kingdom, mainly gas-cooled reactors are employed, operated by the Central Electricity Generating Board. These reactors cause much less occupational exposure than light water cooled reactors (BWRs and PWRs) because there are fewer corrosion products in the primary systems and less stress corrosion cracking of pipes, and steam generator tubing.

The Canadian plants, owned and operated by Ontario Hydro, provide an outstanding example of the success achieved by a large utility (16 operating nuclear power plants) that made large dose reductions to reach the status shown in Figure 3. The Canadian reactors are pressurized heavy-water reactors (PHWRs) of quite different design than reactors with light-water moderation (PWRs) more commonly utilized in the

United States and other countries. The PHWRs produce much more tritium, which is a major additional problem in controlling the dose equivalent with this type of system. Also, the reactor is refueled *during* operation (without shutdown), using a highly sophisticated refueling machine. Early versions of this reactor design, located at the Douglas Point Station, resulted in collective doses about four times higher than U.S. doses per unit of electricity generated during 1967 to 1970. However, through a broad major commitment to improved design for dose equivalent control, the Canadian utility successfully reduced doses from about four person-cSv/MWe-y in 1972 to about 0.3 person-cSv/MWe-y in 1981 (8). Much of this improvement was achieved through elimination from the primary system of alloys containing high cobalt content, addition of shielding, improvements in water purification systems, improvements in air-drying systems (for tritium control), improved reliability and maintainability, and use of fewer, better trained workers. The number of workers per reactor was reduced from about 600 in 1970 to about 300 in 1982. During this same period, the number of workers per U.S. reactor *increased* from about 300 to about 1,100 (2). The Canadian reduction in dose is an outstanding example of what can be achieved by proper design and operation if there is adequate emphasis on dose equivalent control. An important indication of the importance that management placed on the control of dose equivalent at that time was the monetary value used to judge acceptability or break-even in the cost-benefit evaluations. A high value of $10,000 per person-cSv saved was employed, based on the costs of manpower replacement that were being experienced in their plants with high dose equivalent.

These valuations of radiation detriment (in $/person-cSv) include both the objective health detriment (the α term in the ICRP approach to cost-benefit analysis) (24) and "other" detriments (the β term). In nuclear power applications, the β term is generally dominant in the $/person-cSv valuation and therefore it is possible to avoid detailed and controversial evaluations of the health effects costs. The example above illustrates the high valuations of radiation detriment typically employed at nuclear power plants experiencing high collective doses, where the hiring of additional skilled workers is needed to ensure that the work can be done in the time allocated for a shutdown without experiencing overexposures. It is important to realize that during most shutdowns the utility must either start up another facility or purchase replacement power, at a cost of about $20–30,000/h for a 1,000 MWe plant. This cost makes it essential that adequate manpower be available to quickly bring the plants back on line.

Doses at Japanese plants during 1978 to 1982 were high due, in part, to a policy which required more extensive dismantling and testing of components than was required in other countries during annual shutdowns for preventive maintenance. More recently, changes in this policy and other changes have reduced their collective dose equivalents to an even greater extent than for U.S. plants. Some of this reduction may also be attributed to earlier efforts in preventive maintenance.

Doses at French plants are low for several reasons. First, all but a few of the French plants are PWRs that in the United States generally have about half the collective dose equivalent as a BWR of similar size. Second, the French plants are, on average, a few years newer than U.S. PWRs. Because corrosion products build up with time, and maintenance, testing, inspections, and replacement of components all tend to increase with the age of the plants, a significant portion of the difference is due to these factors. Third, the French have standardized plant designs and have more units per station. The standardized plants and multiple units make design and use more cost effective for automatic and remote tooling. Also, their work crews can be trained more thoroughly, and used more effectively than their U.S. counterparts.

Considering the PWR evolution since 1975, for the United States, Japan, and to a smaller extent, for Sweden, Belgium, and the German Federal Republic, the average annual collective dose equivalent per reactor or per unit electricity produced increased at first and then showed a clear decrease (7). In contrast, Switzerland and France show values that are stable or rising slowly.

The results of dose equivalent control in Sweden and Finland have been consistently good (8,9). The BWR plants in these countires are designed by a Swedish steam supplier, ASEA-Atom (now called ABB Atom). These plants have internal recirculation pumps which avoid a major source of exposue due to leaks in the pump seal and they do not require the extra piping that is needed with external pumps. The plant design and layout provide better shielding, better segregation of radioactive components, and they have provisions for adequate work space for routine and special maintenance. Both countries also have very excellent programs for plant chemistry which result in water purity and pH controls that are probably the best in the world. Since water purity has a major influence on the generation of corrosion products, the cracking of pipes, and degradation of components, the careful chemistry is very important. Sweden also has used highly advanced electronic dosimetry to routinely read dosimeters at various work stations, to track dose equivalent by worker, location, specific job or task, and to provide updated information on worker accumulated dose equivalent with respect

TABLE 3—*Factors Which Contribute to Low Doses.*

1. Minimization of cobalt in primary system components exposed to water
2. Careful control of oxygen and pH in the primary system
3. Good primary and secondary system water purity to minimize the formation of corrosion products
4. Careful plant design, layout, and component segregation and shielding
5. Management interest and commitment
6. Minimum number of well-trained workers and in-depth worker training
7. Use of special tools
8. Plant standardization
9. Decontamination of the primary system
10. Pretreatment of surfaces (passivation) to minimize corrosion and deposition

to dose equivalent limits. The success in Sweden may also be due to the goal of 0.2 person-cSv/MWe installed capacity suggested in the 1970's by the National Institute of Radiation Protection. This is equivalent to about 0.3 person-cSv/MWe-y generated, a very ambitious but apparently achievable goal (2).

These are a few of the reasons for the country-to-country differences, but there are others that relate to policies, management practices, special equipment, plant design, and operational practices. It becomes apparent when one begins to probe data for possible causes of plant-to-plant, and country-to-country differences, that there are many causes and, especially in the United States, each plant differs in terms of its mix of important parameters which influence occupational exposure. Several years ago fifteen factors were identified that were important in exposure at nuclear power plants (2). Since then the list was extended to over 75 items, including equipment and organizational items. Some of these are summarized on Table 3.

U.S. Naval Experience (10)

It is interesting to compare the occupational radiation exposure in U.S. Naval Nuclear Propulsion Plants and their support facilities with those of the U.S. nuclear power utilities, since they both have similar problems in maintaining and repairing nuclear reactor systems. Navy personnel were operating 18 tenders, 3 submarine bases, 134 nuclear-powered submarines, and 14 nuclear-powered surface ships at the end of 1987: eight shipyards were engaged in the construction, overhaul, or refueling the ships. Figure 4a,b shows the collective dose equivalent received through 1987 (a) by shipyard personnel and (b) by military and civilian personnel combined in the above facilities, and the total number

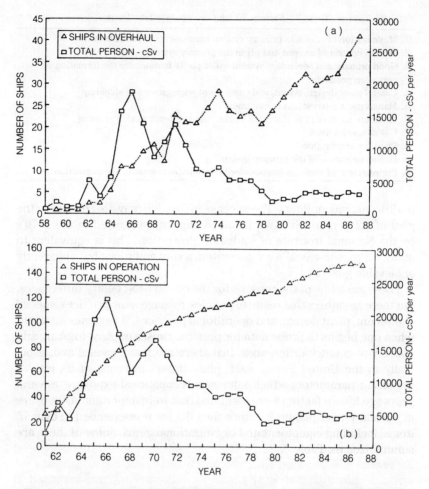

Fig. 4. (a) Collective dose equivalent received by shipyard personnel working in Naval nuclear propulsion plants, 1958-1987.
(b) Collective dose equivalent received by military and civilian personnel in the Naval nuclear propulsion program, 1961-1987.

of ships in operation and in overhaul which increased progressively until 1986. About one fourth of the total collective dose equivalent is received on ships while they are in operation, and approximately three quarters is received by the shipyard personnel who maintain and refuel the vessels in port. The total number of personnel monitored increased from about 6,000 in 1958 to a peak of about 36,000 in 1966, the year that the largest collective dose equivalent was experienced.

The policy of the U.S Navy is to reduce the radiation exposure of personnel to as low a dose as is reasonably achievable. Until 1965, the Navy policy was to use the same limits for radiation exposure as were used throughout the United States, that is 3 cSv per quarter and not to exceed 5(age -18) accumulated dose. In 1965 the International Commission on Radiological Protection (11) reiterated the quarterly and accumulated limits, but suggested that exceeding 5 cSv in one year should be infrequent. In 1967 the Naval Nuclear Propulsion Program took the lead and adopted a rigorous 5 cSv per year limit (10), although the NCRP (12), the ICRP (13), the Atomic Energy Commission (14), and the U.S. Federal Radiation Council (15) were all recommending the 3 c/Sv per quarter, and 5(age -18) accumulated dose limits at that time. As a result of this change, and other emphasis on reducing exposures in the Navy programs, there was a marked decrease in collective dose equivalent per year beginning in 1967 which continued throughout the 1970's until the range of about 50 sieverts per year was achieved; this value has been maintained throughout the 1980's. This achievement is remarkable since the number of ships in overhaul has been increasing markedly, as shown on Figure 4a. Thus, the collective dose equivalent per ship in overhaul, has decreased from about 20 sieverts per year in 1966 to about 1.1 in 1987. This dramatic example of dose equivalent reduction, and the Canadian experience described above illustrate what often can be done if there is a strong effective commitment by management and the necessary funding.

It is remarkable that the number of individuals receiving greater than 5 cSv/y in the Naval programs also dropped from over 500 in 1965 and 1966, to 1 in 1967, and zero thereafter. In addition, the number receiving >4 cSv/y dropped to zero in 1978, >3 cSv/y dropped to zero in 1979, and >2 cSv/y dropped to zero in 1980. In 1987, about 92% of those monitored for radiation in shipyards, and about 99% of those in ships, received <0.5 cSv/y. The average exposure per year for those monitored in ships has been only 0.1 cSv/y. This value is less than the average annual exposure a person receives from natural background and medical radiation, and is because submarines, in particular, are shielded from cosmic radiation and gamma radiation from the ground, which more than compensates for the small dose equivalent received from the reactor onboard. The average lifetime accumulated exposure from radiation associated with Naval nuclear plants for all shipyard personnel is 1.3 cSv; as of 1987, less than 1% received a lifetime exposure >25 cSv.

The ALARA Principle

The ALARA concept has evolved over the past 35 years. While the basic concept has changed little, some interesting variations in termi-

TABLE 4—*Evolution of the ALARA Concept*

Year	Concept	Reference	
1954	As low as *practicable*	NCRP	(12)
1955	*Lowest possible* level	ICRP	(16)
1958	As low as *practical*	NCRP	(17)
1959	As low as *practical*	ICRP	(13)
1960	As low as *practical* and *risk/benefit* balance	FRC	(15)
1965	As low as *readily* achievable; social and economic considerations	ICRP	(11)
1970–76	As low as practicable, $1,000/person-rem	NRC	(20,21)
1973	As low as *reasonably* achievable	ICRP	(19)
1975	Numerical balancing not useful	NCRP	(22)
1977	Optimization	ICRP	(23)
1982	Cost-benefit analysis	ICRP	(24)
1985	ALARA but not quantitative optimization	NRC	(25)
1987	ALARA (economic and social factors)	EPA	(26)

nology and emphasis have been apparent. As shown in Table 4, the concept was expressed in 1954 by the NCRP (12) in terms of "as low as practicable" (ALAP).

The ALAP terminology is still employed in Great Britain and has been used in their legal system for many years. The ICRP's use of the term "lowest possible level" implies a similar goal (16). However, in 1958 and 1959 the NCRP (17) and ICRP(13), respectively, changed the terminology to "as low as practical." Webster (18) states that the terms "practicable" and "practical" are not interchangeable without loss of precision of expression. Practicable applies to something that has not been tested in practice but is presumably achievable, whereas practical stresses opposition to all that is theoretical.

The Federal Radiation Council introduced the concepts of risks and benefits in 1960 (15), and in 1965 the ICRP (11) introduced the term "as low as readily achievable" social and economic considerations being taken into account. Their use of the term *readily* achievable as contrasted to *reasonably* achievable was presumably an error which was corrected in 1973 (19) at which time the use of cost-benefit analysis was suggested. In the early 1970's, the NRC suggested a value which could be used in evaluating off-site exposures to determine if they were as low as practicable. The value suggested was $1,000 per person-cSv (20,21). Shortly thereafter we saw in NCRP Report 43 (22) the statement that numerical balancing is not useful. This statement is in contrast to the ICRP which, in 1977, (23) introduced the concept of optimization of radiation protection as one of the three important elements in their system of dose equivalent limitation. This paper was followed in 1982 by ICRP Publication 37 (24) which dealt in detail with cost-benefit analysis. The Nuclear Regulatory Commission, in its proposed changes to 10CFR20, (25) suggested that ALARA would be required of licensees but not quantitative optimization.

In 1987, the Federal Appellate Court in Washington, D.C. made a ruling that gives additional legal confirmation to the ALARA concept. They stated that the EPA could consider only health and not cost in determining what was a *safe* level of exposure to toxic air pollutants, but that once the safe level was set, the agency could consider costs and other factors in determining how far the polluter must reduce the emissions below this limit (26). Thus, the agency must adopt the standard that first determines the maximum amount of the pollutant above which the risks are unacceptable, and then is required to set an ample margin of safety below that level. For this it may take into consideration costs and technological feasibility.

Application of the ALARA Principle in Nuclear Power Plants

In light of the relatively high collective doses being received at U.S. nuclear power plants compared to most other countries, and compared to U.S. naval reactor facilities, the question to be answered is—Are the U.S. plants operating in a manner consistent with the ALARA principle? The regulations and regulatory guides of the Nuclear Regulatory Commission (25,27,28) have consistently stated that nuclear power plants should be designed and operated in accordance with the ALARA principle. Similarly, the utilities' self-regulating organization, INPO, also has urged that plants practice ALARA concepts. Most U.S. plants now have ALARA committees at corporate and plant levels which aid in formulating and carrying our ALARA policies and promote effective communications and working relations among the various groups that must interact to make these policies effective. In addition, it is common practice to track doses for high dose equivalent jobs, to plan and implement dose equivalent control practices for these jobs, and to conduct post-job reviews to summarize lessons learned and provide documentation for future planning.

It has become common practice in the United States to employ a specific dollar per person-cSv value to judge the cost effectiveness or break-even point for various dose-equivalent control or dose-equivalent reductions and modifications. Often, the application of the cost-effectiveness criterion is first considered crudely in terms of total capital investment and total annual operating cost or savings over the remaining life of the plant, or the life of the equipment being considered, and total collective dose equivalent saved over the same period. The ratio of these two values (net costs per cSv saved) gives a value of cost effectiveness, that can be compared to the valuation of detriment employed at the plant. The valuation of detriment is generally based upon a consideration of the number of workers near the dose equivalent limit and, therefore, how many additional workers may be needed in future to do the various high dose equivalent jobs. If one additional worker may need to be hired, the salary, fringe benefits, and costs of training, can be compared to the additional collective dose equivalent that is made available to the collective dose equivalent pool. This ratio gives one value of dollar per person-cSv. Typical values in U.S. nuclear utilities today range from $1,000 to $10,000 per person-cSv.

Table 5 gives capital costs, cost-effectiveness of dose equivalent saved, and other related values for some typical modifications of dose control (29,30). The table illustrates that considerations, which go far beyond the simple cost-effectiveness ratio, usually enter into the decision-making process. For example, both monetary costs and savings of collective dose equivalent are considered less valuable in future years than those which will occur during the present year due to the changes in time in the value of money. The usual procedure is to discount all future costs and savings by, *e.g.*, 4% per year (the typical uninflated discount rate), and thereby convert all future costs into present values.

Table 5 illustrates the cost effectiveness for several modifications with positive net costs, positive net dose equivalent savings, and a cost-effectiveness ratio that is favorable (29,30). The items have been listed in decreasing order of cost effectiveness, as shown in the second column, and are based on a "basic" model in which future costs and savings are not converted to present worth values by discounting. Thus, if cost effectiveness is used as the sole criterion, the most cost effective modification would be the first one listed in column 2. A similar ranking of results is obtained if future costs and savings are discounted by 4% as shown by values given in column 3.

However, a utility is also concerned with profits expected in proportion or in comparison to dollars initially invested. This would be a benefit/cost ratio that considers both dose equivalent savings and future operational savings. In this case, one must apply a value of ($/person-cSv) to dose equivalent savings to convert them into monetary equivalents. Using $1,000/person-cSv yields values shown in the 4th column. The data illustrate that this benefit/cost ratio would lead to the selection of either the first two or the last item as being mostly beneficial.

TABLE 5—*Comparisons of Various Decision Criteria*

Project	Basic Model Cost-Effectiveness ($/Person-cSv)	Cost-Effectiveness ($/Person-cSv)*	Benefit/Cost ($Saved/$Invested)*	K$ Invested	cSv Saved (Disc.)**	Discounted cSv Saved/ K$ Invested)*
BWR-CRD Hydrolazing	20	35	27	5.9	300(170)	29
SG Head, Portable Shield	49	86	22	35	1,500(850)	24
CVCS Shields	59	100	9.0	1.8	30(17)	9.4
Clean Seal Water	62	110	9.2	33	600(340)	10
PWR Level Monitor, N-16	68	120	7.4	16	240(140)	8.8
Low Cobalt Coolant Pumps	62	120	7.5	35	560(300)	8.6
Low Cobalt CRD Mechanism	76	140	6.3	59	810(430)	7.3
Low Cobalt in SG Tubes	75	140	6.3	349	4,700(2,500)	7.2
Reactor Head Shielding	170	180	24	1.9	88(55)	29

*Based on present worth values discounted at 4%/year. $ saved calculated using $100,000/person-Sv ($1,000/person-rem).
**Values in parentheses are discounted at 4%/year.

Column 5 shows the total capital that must be invested for each of these modifications, another criterion which often influences the decision process. Because capital budgets are always limited, one is frequently forced to select options which have small capital investments, although they may not be optimum in terms of cost effectiveness or dose equivalent savings. In this case, the third and last items are most desirable because they have very small capital investment associated with them.

For a plant which has been experiencing very high doses over a period of years, pressures from the NRC, INPO, plant workers, plant management, and the general public, may lead to them giving a high priority to the total dose equivalent saved, in which case, the values listed in column 6 become more important. For long-term benefits in terms of total collective dose equivalent saved, the next to the last item would be most beneficial, which is replacement of a steam generator with one having low concentrations of cobalt in the steam generator tubes. In this example, replacement of the steam generator was required for other reasons. Therefore, the costs and benefits reflect differentials expected for a specification of low cobalt in the steam generator tubes (compared to normal specifications). The capital costs for this item are relatively high and may require longer term planning, scheduling, and approvals by various levels of management, and perhaps even by the state board which governs electricity utility rates. The values shown in parentheses in column 6 illustrate the effect of discounting future dose equivalent savings. Although few utilities discount future dose equivalent savings at present, a logical case can be made for doing so since the costs of these future doses are dominated by the costs of replacing workers. Those costs can properly be treated as other operational costs and therefore, be discounted. If the valuation of detriment was made purely on the basis of health effects, then the discounting of future doses becomes somewhat more debatable, though a case can still be made for doing so.

Another criterion which may be employed is the ratio of discounted dose equivalent saved compared to capital investment (column 7, Table 5). In this case the same selections would be made as for the benefit/cost ratio criterion shown in column 4. This correspondence is coincidental and relates to the fact that operational dollars saved and dose equivalent saved tend to correlate strongly since many of the operational dollar savings are due to reduced manpower requirements.

Finally, one could calculate total dollars saved based on a cost-benefit calculation in which the detriment is valued at $1,000 per person-cSv.

Using this criterion, one again would obtain the same ranking as given in column 6.

Other factors frequently employed in a cost-benefit or cost-effectiveness evaluation include state, local, and federal taxes, and the total impact of costs on annual expenditures (or cash flow) in inflated dollars because this affects the rate increases a utility may need to request and also affects its annual operating budget.

These examples illustrate why cost-benefit analyses and cost-effectiveness analyses are considered aids in the decision-making process, but often are not decisive factors. It is improtant, however, that the cost-benefit and cost-effectivenes sprocesses be developed and implemented because they lead to a consistent decision-making process and provide important information useful for prioritizing actions and judging the quantitative part of the ALARA process.

Conclusions

We have drawn several conclusions from this discussion.

1. Collective occupational dose equivalents in the nuclear power industry in the United States are higher per plant, and per unit electricity generated, than in most other developed countries.
2. Very few workers are exposed beyond the present regulatory limits.
3. Wide differences exist between collective dose equivalents per plant from country to country, and also from plant to plant within the United States.
4. Large reductions in exposure were achieved in Canadian utilities and in the U.S. Naval Nuclear Propulsion Programs.
5. Implementation of quantitative aspects of the ALARA principle are complex, and require additional development and application.
6. Plant and national goals for dose minimization are important.

The number and variety of high dose equivalent jobs in nuclear power plant operation and maintenance make it important to use a systematic and comprehensive approach to ALARA. Because of the interdependence of many aspects of plant operations, and their impact on occupational dose equivalent it is imperative that evaluations include considerations not only of the effects of individual modifications but also of the impact of other modifications that may alter the dosimetric and cost impacts of a given modification. For example, in PWRs work on steam generators is the major source of occupational exposure. These exposures can be reduced: by improving plant chemistry (*e.g.* operating at an optimum pH); by improved access and work space for maintenance activities to make them more efficient; by improved shielding around other components in the area of steam generator maintenance to avoid dose equivalent from components not being worked on; by decontamination of primary systems before work begins; and, by use of remote and automatic maintenance devices. The relative importance and cost effectiveness of the remote devices, for example, depends upon how much has been invested in each of the former devices for dose equivalent control. Similarly, the value of system decontamination depends upon whether or not remote tooling is already available at the plant, or may become available in the future.

Much has been learned through the operation and maintenance of the current generation of nuclear power plants which is being used to provide design information for the next generation of plants: these are predicted to have much lower occupational exposures—50 person-cSv/y for the advanced BWR (31) and 100 person-cSv/y for PWRs (32). The present challenge in radiation protection in the nuclear power industry is to find the optimum level of radiation protection, which depends upon both quantitative ALARA cost-benefit considerations, and less easily quantified evaluations of the acceptance of radiation exposures by workers and society.

References

1. *World Nuclear Industry Handbook*, Supplement to Nuclear Engineering International Magazine (Nuclear Engineering International, Quadrant House, The Quadrant, Sutton, Surrey SM2 5AS, United Kingdom) (1989).
2. Baum, J.W. and Horan, J.R. *Summary of Comparative Assessment of U.S. and Foreign Nuclear Power Plant Dose Experience,* Report NUREG/CR-4381, BNL-NUREG-51918 (U.S. Nuclear Regulatory Commission, Washington) (1985).
3. Kumazawa, S., Nelson, D.R. and Richardson, A.C.B. *Occupational Exposure to Ionizing Radiation in the United States—A Comprehensive Review for the Year 1980 and a Summary of Trends for the Years 1960–1985,* Report EPA 520/1-84-005 (Office of Radiation Programs, U.S. Environmental Protection Agency, Washington) (1984).
4. Brooks, B.G. *Occupational Radiation Exposure at Commercial Nuclear Power Reactors and Other Facilities 1985,* Eighteenth Annual Report, Report NUREG-0713, Vol. 7 (U.S. Nuclear Regulatory Commission, Washington) (1988).

5. INPO. Institute of Nuclear Power Operations *1988 Performance Indicators for the U.S. Nuclear Utility Industry* (Institute of Nuclear Power Operations, Atlanta, Georgia) (1989).
6. Khan, K.A. and Baum, J.W. *Worldwide Activities on the Reduction of Occupational Exposure at Nuclear Power Plants*, Report NUREG/CR-5158, BNL-NUREG-52086, Vol. 1 (Nuclear Regulatory Commission, Washington) (1988).
7. Lochard, J. and Benedittini, M. *Expositions Professionnelles dans les Réacteurs à Eau Pressurisée: Comparaison Internationale de Quelques Indicateurs Globaux entre 1975 et 1985*, Report 103 (Centre d'etude sur l'évaluation de la Protection dans le Domaine Nucléaire, 92260 Fontenay-aux-Roses, France) (1987).
8. Wilson, R., Chase, W.J. and Sennenia, L.J. "Occupational dose reduction experience in the Ontario Hydro Nuclear Power Stations," Nuclear Technology, **72,** 590 (1986).
9. Horan, J.R., Baum, J.W. and Dionne, B.J. *Proceedings of an International Workshop on Historic Dose Experience and Dose Reduction (ALARA) at Nuclear Power Plants*, Report NUREG/CP-0066, BNL-NUREG-51901 (Nuclear Regulatory Commission, Washington) (1985).
10. Mangeno, J.J. and Tryon, A.E. *Occupational Radiation Exposure from U.S. Naval Nuclear Propulsion Plants and Their Support Facilities*, Report NT-88-2 (Department of Navy, Washington) (1988).
11. ICRP. International Commission on Radiological Protection. *Recommendations of the International Commission on Radiological Protection*, ICRP Publication 9 (Pergamon Press, Oxford) (1966).
12. NCRP. National Council on Radiation Protection and Measurements. *Permissible Dose from External Sources of Ionizing Radiation (1954) including Maximum Permissible Exposure to Man*, Addendum to National Bureau of Standards Handbook 59, NCRP Report No. 17, p. 20–22 (National Council on Radiation Protection and Measurements, Bethesda, Maryland) (1954).
13. ICRP. International Commission on Radiological Protection. *Recommendations of the International Commission on Radiological Protection*, ICRP Publication 1 (Pergamon Press, Oxford) (1959).
14. Code of Federal Regulations Title 10 (Energy) Part 20, "Standards for protection against radiation," Federal Register, 22 FR 548 (U.S. Government Printing Office, Washington) (1957).
15. FRC. Federal Radiation Council. "Radiation protection guidance for federal agencies," Federal Register, 25 FR 44031 (U.S. Government Printing Office, Washington) (1960).
16. ICRP. International Commission on Radiological Protection. "Recommendations of the International Commission on Radiological Protection," British J. Radiol., Supplement 6 (1955).
17. NCRP. National Council on Radiation Protection and Measurements. "Maximum permissible radiation exposures to man," NBS Tech. News Bull. **41,** 17 (1957).

18. Webster's New Collegiate Dictionary (G. & C. Merriam Co., Springfield, Massachusetts) (1976).
19. ICRP. International Commission on Radiological Protection. *Implications of Commission Recommendations that Doses be Kept as Low as Readily Achievable*, ICRP Publication 22 (Pergamon Press, Oxford) (1973).
20. U.S. Atomic Energy Commission. "Light-water-cooled nuclear power reactors," Federal Register, 35 FR 18385 (U.S. Government Printing Office, Washington) (1971).
21. NRC. U.S. Nuclear Regulatory Commission. *Calculating of Annual Doses to Man from Routine Releases of Reactor Effluents for the Purpose of Evaluating Compliance with 10 CFR Part 50, Appendix I*, NRC Regulatory Guide 1.109; superseded by Revision I, dated November 1977 (U.S. Nuclear Regulatory Commission, Washington) (1976).
22. NCRP. National Council on Radiation Protection and Measurements. *Review of the Current State of Radiation Protection Philosophy*, NCRP Report No. 43 (National Council on Radiation Protection and Measurements, Bethesda, Maryland) (1975).
23. ICRP. International Commission on Radiological Protection. *Recommendations of the International Commission on Radiological Protection*, ICRP Publication 26; Annals of the ICRP **1,** 3 (Pergamon Press, Oxford) (1977).
24. ICRP. International Commission on Radiological Protection. "Cost-benefit analysis in the optimization of radiation protection." ICRP Publication 37 (Pergamon Press, Oxford) (1982).
25. NRC. U.S. Nuclear Regulatory Commission. "Proposed Rules 10 CFR Parts 19, 20, 30, 31, 32, 34, 40, 50, 61, and 70," Federal Register, 50 FR 51992 (U.S. Government Printing Office, Washington) (1985).
26. Fishbein, G.W. (Ed.) *Environmental Health Letter* **26,** No. 15 (1987).
27. NRC. U.S. Nuclear Regulatory Commission. *Information Relevant to Ensuring that Occupational Radiation Exposures at Nuclear Power Stations will be as Low as is Reasonably Achievable*, NRC Regulatory Guide 8.8, Revision 2 (U.S. Nuclear Regulatory Commission, Washington) (1977).
28. NRC. U.S. Nuclear Regulatory Commission. *Operating Philosophy for Maintaining Occupational Radiation Exposures as Low as is Reasonably Achievable*, NRC Regulatory Guide 8.10, Revision 1-R (U.S. Nuclear Regulatory Commission, Washington) (1977).
29. Baum, J.W. and Matthews, G.R. *Compendium of Cost-Effectiveness Evaluations of Modifications for Dose Reduction at Nuclear Power Plants*, Report NUREG/CR-4373, BNL-NUREG-51915 (U.S. Nuclear Regulatory Commission, Washington) (1985).
30. Baum, J.W. "Overview, What is ALARA? Are We There?," in *Proceedings of the Twenty-first Midyear Topical Meeting*, Health Physics Society, Bal Harbour, Florida (1987).

31. Wilkins, D.R., T. Seko, S. Sugino and H. Hashimoto, "Advanced BWR: Design improvements build on proven technology," Nuclear Engineering International **31,** 36 (1986).
32. Iacovino, J., "Dose reduction: advanced PWR aims at 100 man-rem/year," Nuclear Engineering International **30,** 48, (1985).

Status and Trends in Radiation Protection and Medical Workers

William Hendee
American Medical Association
Chicago, IL

F. Marc Edwards
St. Luke's Hospital of Kansas City
Kansas City, MO

Abstract

   Medical workers constitute the largest group of occupationally exposed workers and receive the second largest collective effective dose equivalent, exceeded only by nuclear fuel cycle workers. Although the number of medical workers has risen steadily over the past three decades, their collective dose has steadily declined. Significant factors effecting the patterns of medical occupational exposure are the increased demand for medical imaging, changing patterns of the practice of radiology and the development of new imaging technologies. Although there have been recent societal demands to interpret personnel monitoring results in terms of risk, this goal is particularly difficult for medical workers. Problems of monitoring device placement, inconsistent utilization of shielding, and conversion of limited monitoring data to effective whole body dose remain to be solved. Emerging trends such as increased utilization of non-hospital based imaging centers and the implementation of non-medical criteria in the management and delivery of health care may have a future impact on medical radiation protection.

Introduction

Occupational exposure to radiation in medicine is the direct result of the practice of medicine. Trends in medical radiation exposure and protection are affected not only by developments in health physics, but also by those in the practice of medicine. The collective dose equivalent to occupationally exposed individuals in medicine has been steadily decreasing over the past twenty years, while the utilization of radiation in medicine has been steadily rising. Does this decrease in occupational exposure reflect improvements in technology and radiation protection or is it due to changes in the practice of medicine? And can we expect these trends to continue?

Characteristics of Medical Occupationally Exposed Workers

What is meant by a medical occupationally exposed worker? The term "medical worker" is often used informally to denote employees of hospitals, clinics and private offices who are involved in the delivery of health care to humans. This definition tacitly excludes dental, veterinary and other users of "clinical" radiation. The EPA has adopted a broader definition of "medical worker" that includes hospital, clinic, private office medical and dental employees as well as podiatry, chiropractic and veterinary employees (1). The EPA further distinguishes "potentially exposed" employees from "exposed" employees, with the latter defined as workers who receive a minimum detectable exposure or more during a one-year period. By tabulating potentially exposed workers, the EPA seeks to identify persons who work around radiation sources but are not individually monitored. Under the EPA definition, there were approximately 584,000 potentially exposed medical occupational workers in the U.S. in 1980. Of those, approximately 306,000 were estimated to have received less than a minimum detectable exposure.

The EPA classification of medical workers separated by job category, mean annual whole-body dose and collective dose for 1980 is shown in Table 1 (1). Contributions from podiatry, chiropractic and veterinary practices account for less than five percent of the medical collective dose. Hospital, private practice and dental workers constitute the greatest numbers of occupationally exposed workers. Primarily because of their large numbers, these groups contribute the largest fraction of the medical collective radiation dose. Dental workers, whose numbers are as great as hospital and private practice workers combined, received substantially less exposure and account for only 14 percent of the medical collective dose.

TABLE 1—*National occupational exposure summary of the medical sub-group for 1980 as found by Kumazawa et al., (1)*

Occupational subgroup	Number of Workers[a]		Mean annual whole-body dose (mSv)[d]		Annual Collective dose[a,d]
	Total[b]	Exposed[c]	Total	Exposed	(person-Sv)
MEDICINE					
Hospital	126,000	86,000	1.40	2.00	172
Private Practice	155,000	87,000	1.00	1.80	160
Dental	259,000	82,000	0.20	0.70	56
Podiatry	8,000	3,000	0.10	0.30	1
Chiropractic	15,000	6,000	0.30	0.80	5
Veterinary	21,000	12,000	0.60	1.10	13
ENTIRE SUBGROUP	584,000	276,000	0.70	1.50	407

[a]Numbers of workers are estimated values and are rounded to nearest thousand. Mean doses are rounded to the nearest 0.1 mSv, and collective doses to the nearest person-Sv.
[b]All monitored and unmonitored workers with potential occupational exposure.
[c]Workers who received a measurable dose in any monitoring period during the year.
[d]To convert to mrem and person-rem, multiply respective figures by 100 (1 mSv = 100 mrem, 1 person-Sv = 100 person-rem).

An additional important characteristic of occupational exposure is the frequency distribution of dose equivalent and collective dose equivalent. Data for the total medical workforce, including all potentially exposed workers both monitored and unmonitored, are shown in Table 2 (1). These data, like most other occupational dose distributions, illustrate several typical features. First, the distribution is heavily skewed towards low annual dose equivalents. Although the average dose equivalent was about 0.7 mSv/y (70 mrem/y) for the total potentially exposed workforce, approximately 53 percent of the group received less than measurable exposure and 88 percent received less than 1 mSv/y (100 mrem/y). Of the exposed workers (those receiving a measurable dose equivalent at least once during the year), the average dose equivalent was 1.5 mSv/y (150 mrem/y). Still, approximately 75 percent received less than 1 mSv/y (100 mrem/y). Second, a large contribution to collective dose was made by the small fraction of workers who received relatively high dose equivalents. The three percent of the total workforce (six percent of the exposed group) who received a dose greater than 5 mSv/y (500 mrem/y) contributed approximately 51 percent of the collective dose. Finally, less than 0.5 percent of the total workforce received a dose equivalent in excess of 20 mSv/y (2 rem/y) and less than 0.05 percent received a dose equivalent in excess of the recommended upper occupational limit of 50 mSv/y (5 rem/y).

The distribution of medical occupational exposures over time is poorly understood. It would be interesting to know, for instance, if individuals

TABLE 2—*Dose equivalent and collective dose frequency distribution by dose range for the 1980 medical radiation workforce*[a].

Dose equivalent range (mSv/y)	Number of workers	Percent of workers	Collective dose (Person-Sv)	Percent of collective dose
0–MD	306506	52.5	8.50	2.1
MD–1	205969	35.3	68.43	16.8
1–2.5	36078	6.2	61.21	15.0
2.5–5.0	18284	3.1	59.52	14.6
5–10	10324	1.8	66.55	16.4
10–20	4629	0.79	65.62	16.1
20–30	1321	0.23	32.73	8.0
30–40	467	0.08	18.36	4.5
40–50	217	0.004	9.85	2.4
50–80	205	0.004	12.69	3.1
80–120	40	0.0007	3.54	0.9
120+	0	0	0	0

[a]The data include both exposed and potentially exposed workers for all occupational subgroups given in Table 1. A large portion of the workers in the 0–MD (Minimum Detectable) range are dental workers. Each occupational subgroup follows the same type of distribution but with a different mean (1). To convert to mrem and person-rem, multiply respective figure by 100.

receiving high readings in one year repeat that behavior in succeeding years. There is some reason to suspect that this may be the case, because subspecializaton of job tasks can lead to the same group of workers performing the same tasks year after year. For example, fluoroscopy is the radiologic procedure most responsible for medical occupational exposure. If a single subgroup of medical workers is consistently involved in fluoroscopy, then repetitious high exposures would be expected for this subgroup. On the other hand, rotation of workers through all tasks in a radiology department would tend to even the time distribution of occupational exposure to these workers. Objective data are inadequate at present to resolve this issue. However, patterns of work performance in radiology facilities suggest that subgroups tend to maintain their identity, and therefore, their exposure patterns to radiation from year to year.

Medical workers can be characterized by the types of institutions in which they work. For the most part, health care is delivered by institutions that would individually not be considered "large businesses". Medical radiation workers are rather dispersed with comparatively few workers per institution. The average radiology practice has approximately five radiologists, for example, and only two percent of practices have more than ten radiologists. The EPA has estimated that each radiologist may account for ten workers (1). A five hundred bed hospital may monitor from 50 to 150 employees, but only six percent of U.S. hospitals have 500 or more beds. A small hospital, clinic or private office is likely to have fewer than fifteen monitored employees. In these institutions the responsibility for radiation protection of medical workers is likely to reside in an individual whose training and duties are not full time professional radiation protection. If a physicist is available, the duties of this individual may not significantly involve radiation protection. In a recent manpower survey of imaging scientists by the American College of Radiology, 89 percent of respondees worked in a hospital or medical school, yet most spent less than 15 percent of their time on radiation protection activities (2). In these situations one wonders "who is minding the store?"

Historical Trends

Past trends in medical occupational exposure have been studied by the EPA(1). Reasonably good data exist for the period 1965 to 1980. These data can be combined with utilization data from other investigators to yield a reasonably good overview of radiation exposure conditions over this 15 year period (3, 4, 5, 6). Shown in figures 1 and 2 are the relative number of procedures, the relative number of potentially exposed workers and the relative collective dose for the years 1965-1980. These data have been normalized to their values in 1965 to reveal trends over the succeeding 15 years.

Trends in dental exposure are shown in Figure 1. The number of procedures and workers increased linearly at a rate of roughly five percent per year over the period 1965–80. Over the same interval, the collective dose to dental workers decreased by about 3.5 percent per year. In 1980 the collective dose from dentistry was approximately half that in 1965. Occupational dose is related to patient dose, so changes in dental patient exposures are of interest during this period. Figure 1 shows the mean exposure at skin entrance for a dental bitewing radiograph (7), multiplied by the relative number of procedures. This product relates directly to patient collective dose. Most dental imaging is bitewing radiography, and most occupational exposure is received from this procedure. Thus it is not surprising that occupational and patient exposures follow the same trend. The decrease to 1/4 of the bitewing entrance exposure from 1965 to 1980 has been attributed to the use of faster film, tighter beam collimation and improved quality control(7). Even though dental workers comprised 44 percent of all medical workers in 1980,

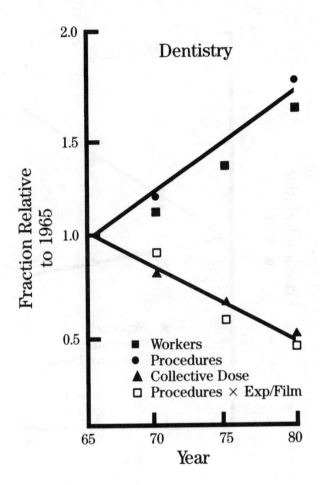

Fig. 1.

they accounted for only 14 percent of the medical collective dose. The remaining discussion focuses primarily on medical (*i.e.* non-dental) occupational exposure.

Trends in hospital/private practice occupational exposures are shown in Figure 2. Here also, the number of procedures and workers increased linearly at a rate of roughly five percent per year. The collective dose of the workers decreased by about 1.5 percent per year, so that the collective dose in 1980 was approximately 80 percent that in 1965. The collective dose of hospital/private practice workers was estimated in 1980 to be 330 person-Sv (33,000 person-rem). Data concerning effective whole body collective dose equivalent to patients are not available for this time

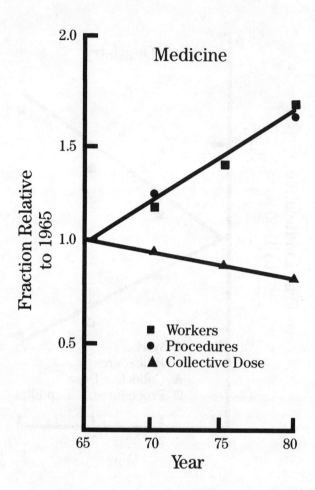

Fig. 2.

period. The NCRP has estimated that in 1980 the annual collective effective dose equivalent was 123,000 person-Sv (123,000,000 person-rem) to the U.S. population due to diagnostic radiology and nuclear medicine (8). The mean entrance exposure for chest, abdomen and spine radiographs decreased by approximately 20 to 25 percent from 1965 to 1980 (7), while the imaging volume increased approximately 65 percent. Similar data are not available for fluoroscopic and nuclear medicine procedures.

The picture that emerges from these trends can be summarized as slowly decreasing collective dose to occupationally exposed individuals in the face of increased utilization of x rays. Since the occupational collective dose has decreased, the observed reduction in individual average annual exposure has not been achieved by a simple increase in the number of employees. Individual dose limits are rarely (less than one percent of workers) exceeded in medicine. Indeed if the only objective of radiation protection was to assure that the annual dose limit of 50 mSv (5000 mrem) was satisfied, occupational radiation protection in medicine would be considered an unqualified success. It is more difficult to assess whether medical occupational exposure is as low as reasonably achievable (ALARA). The trend of decreasing collective dose is at least circumstantial evidence that additional reductions in exposure might be achievable. Whether or not these reductions are warranted under the ALARA criteria is a separate issue. To understand past trends and to predict those in the future, factors that influence medical occupational exposure must be considered in greater detail.

A Simple Model

To analyze trends in radiation exposures one must first identify the major factors that can impact upon these exposures. While these factors are complex in the case of medical occupational exposures, it is possible to build a simplified model as shown in Figure 3. We start with a general population N, of individuals, such as the population of the United States. Since the driving force behind the use of radiation in medicine is the diagnosis and treatment of disease, the first question to be asked is what fraction per year of the population are patients. This fraction is denoted by f_p. As either the size of the population increases or f_p increases, the absolute number of patients increases and, all other variables being held constant, the medical occupational exposure would be expected to increase. The population of the U.S. has been growing at a fairly constant rate. In addition, many factors affect f_p, including aging of the population, epidemics, lifestyle and healthcare patterns, dietary practices, and the introduction of preventive measures.

The next factor to consider is the fraction of patients receiving medical imaging. This fraction is denoted as f_i. We could also include in this fraction the number who receive treatment with radiation producing devices. This population of patients is very small, however, exposures are administered routinely under carefully supervised conditions. Hence, radiation therapy does not contribute substantially to medical occupational exposure, and the model is confined solely to imaging.

The utilization of medical imaging is determined primarily by medical indications. These indications should be directly related to medical

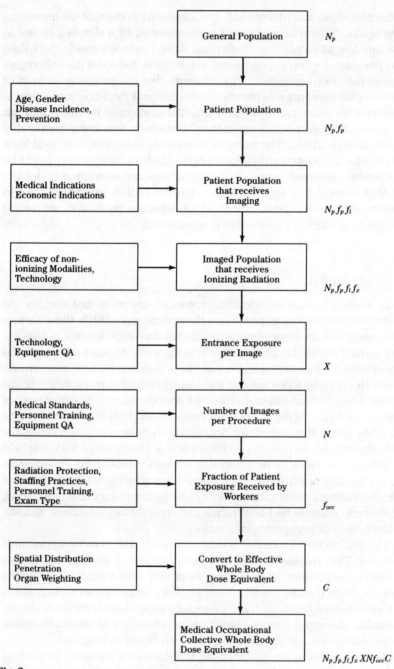

Fig. 3.

efficacy, so that patients do not receive imaging procedures unless they are beneficial. However, examples are not infrequent where examinations of questionable benefit are administered to the patient. Professional and governmental organizations and journals continue to address the issue of utilization through refinement and promulgation of clinical indications for the use of medical imaging.

In addition to medical indications for imaging, there may be additional considerations that influence the decision. One such consideration may be economic, such as the recent enactment of "diagnostic related group" (DRG) reimbursement. Many believe that this economic policy has decreased the utilization of imaging, although no definitive studies have been performed. Another "non-medical" indication is the practice of "defensive" medicine out of fear of legal action. In this case utilization is increased as physicians order medically unnecessary examinations to avoid the appearance of having conducted an incomplete evaluation of the patient.

Having determined the fraction of patients who receive an imaging procedure, we next must identify the fraction of these procedures that utilize ionizing radiation. This fraction is denoted f_x and documents the possible competition between ionizing and non-ionizing imaging procedures. Ultrasound, magnetic resonance imaging (MRI) and endoscopy are currently the most prevalent competitors to imaging procedures employing ionizing radiation. The major factors influencing choices among these imaging procedures are their diagnostic accuracy and medical efficacy. Other important factors include the availability and cost of equipment, the forcefulness of advocates for the different technologies, and the ability of physicians to interpret the images, particularly in the case of a new technology.

In a more sophisticated model, the flow diagram would now branch into each possible imaging procedure employing radiation. To simplify the model, we will assume that there is only one procedure. It is now necessary to know how much radiation is utilized to perform the procedure. Because this analysis focuses on exposure to the worker and not to the patient, a fairly unrefined measure, such as the entrance exposure per image, is a sufficient descriptor. The total radiation utilized for a procedure is simply the product of the entrance exposure per image, denoted by X, and the number of images per procedure, denoted by N. X depends upon the design of the equipment, especially the sensitivity of the image receptor, and on the state in which the equipment is maintained. A well run quality assurance program has been shown to reduce patient exposure (9). The number of images per procedure depends upon medical standards, how well the personnel performing the examination execute the procedure and the state of equipment maintenance. The last two factors influence radiation dose principally through the "repeat rate".

Next, the fraction of patient exposure received by the medical occupational work force must be estimated. This fraction is denoted f_{occ}. In an overly simplistic example where a single radiation worker is positioned without protective garments next to a patient during an imaging procedure, f_{occ} would be just the fraction of radiation scattered from the patient to the worker. In actuality, f_{occ} depends upon the type of examination, the radiation protection measures taken by the workers, how many workers are present during the procedure and so forth. In a general sense, f_{occ} is the fraction of patient collective dose that is "converted" to medical occupational collective dose. Its use allows one to explicitly delineate the medical occupational exposure that results from patient exposure.

Finally, medical occupational exposure must be converted into a meaningful dose equivalent. This conversion requires consideration of the spatial distribution of the radiation field, the penetrability of the radiation within the worker, and the adjustment of non-uniform organ doses to an effective whole body dose equivalent (DE). The conversion factor, denoted as c, is discussed below. It must be included to reveal the problems associated with risk assessment evolving from simple estimates of exposure.

With the caveat that the model described is relatively simplistic, we can now formulate an expression for the collective whole body effective dose equivalent of medical occupationally exposed individuals. This expression is:

$$DE = N_p f_p f_i f_x X N f_{occ} C$$

Only one of the factors in this equation, namely f_{occ}, is directly controllable by the personnel responsible for radiation protection in the medical environment.

With this simple model in mind, we shall now consider the trends from 1965–1980 that were reviewed earlier. It is apparent that the product $N_p f_p f_i f_x$, the number of patients imaged with ionizing radiation, steadily increased over that time period. While the population increased by only about 20%, the number of medical x ray procedures increased approximately 65% and the number of dental x ray procedures increased 87%. Mettler has pointed out that in 1970, 15% of medical diagnostic x ray examinations were performed on the 10% of the population classified as elderly (greater than 65 years of age) (5). By 1980 over 25% of hospital exams were performed on this group, yet it had grown to only 11% of the population. He speculates that the increased utilization of medical imaging procedures for the elderly had more to do with increased reimbursement provided by Medicare than with greater incidence of disease among the elderly (5). It should be noted that over most of this period non-ionizing imaging procedures had small impact. For example, ultrasound constituted less that 0.5% of medical imaging studies in 1973 and had grown to only 4% by 1980 (6).

Procedures exerting the greatest influence on medical diagnostic occupational exposure are those that include fluoroscopy. These procedures, primarily examinations of the upper and lower gastrointestinal tract (UGI and BE), special vascular procedures and cardiac catheterizations, constitute the greatest sources of worker radiation exposure. From 1965 to 1980 the number of UGI and BE exams increased 47%, and from 1973 to 1980 the number of special vascular and cardiac catheterization exams increased over 100% (6, 5). By 1980 approximately 8% (14 million) of medical diagnostic imaging procedures involved fluoroscopy. While the rate of increase of fluoroscopic procedures was less than that of imaging overall, it still exceeded the growth rate of the population over the same interval.

In the light of increases in medical radiation examinations, especially fluoroscopy, what accounts for the decrease in medical occupational collective dose? In the case of dental exposure, most of the credit goes to the increased sensitivity of the image receptor. For diagnostic radiology the answer is less clear. Possibilities include improvements in fluoroscopic image receptor sensitivity, decreased fluoroscopic imaging time per procedure and improvements in radiation protection practices. From 1965 to 1980 fluoroscopic equipment improved dramatically as televised image-intensification became widely available, followed by introduction of the cesium iodide image intensifier that absorbs radiation about three times more efficiently than does zinc cadmium sulfide. Federal requirements on fluoroscopic beam limitation, table top exposure rates, audible timers and shielding drapery were also instituted during this period. One could also postulate that improvements in image quality may have reduced the imaging time required to reach a diagnosis. The impact of radiation protection practices is difficult to evaluate. One might postulate that improved instruction of diagnostic radiology residents and radiologic technologists led to increased awareness of radiation protection. Federal and state requirements regulating the use of radiation and the qualifications of users also could have had an impact.

Future Possibilities

Trends in medical radiation procedures can be summarized as a dramatic increase in utilization, with the possible increase in occupational exposure neutralized and even reduced through technological developments, particularly more sensitive image receptors. To what extent can we expect these trends to continue?

Future trends in utilization will probably be dominated by conflicting forces. On the one hand, the aging of the population, coupled with ever increasing expectations of higher quality health care, will build demand for greater use of medical imaging. Health care will probably continue to become increasingly outpatient oriented, and more imaging will be carried out in free standing imaging centers that are often investor owned. Some observers have speculated that changing patterns of the financial structure of the medical industry could produce pressure to over-utilize profitable procedures. Continued emphasis on "defensive medicine" in response to fear of malpractice suits could also increase medically unnecessary radiologic procedures.

On the other hand, these trends of increased utilization are opposed by the ability and willingness of society to pay the bill. A major effort has been launched to control the cost of Medicare expenditures through the DRG reimbursement process for hospital-based procedures. A 1985 survey of radiology managers detected a decrease in the number of diagnostic imaging procedures performed in the hospital (11). If this limited study is valid, it is the first time since 1965 that a decrease in hospital utilization of imaging has occurred. The author of the study blamed the decrease on technological innovations, new reimbursement policies, and probably a shift to outpatient imaging. In addition to Medicare changes, the private sector is attempting to control costs through health maintenance organizations (HMO) and preferred provider organizations (PPO). Such organizations often try to control utilization as a means of controlling cost.

Even if the utilization of imaging continues to increase, patient and occupational exposures could decrease because of developments in radiation imaging modalities and shifts to procedures that do not employ ionizing radiation. Ultrasonic imaging has experienced very rapid growth, particularly since the late 1970's. Endoscopic imaging has also expanded, particularly with the development of flexible televised endoscopes with convenient hard-copy documentation. Magnetic resonance imaging is just now becoming widely available. Of particular interest are new nonionizing imaging procedures that replace fluoroscopic procedures. Thus, from the viewpoint of occupational exposure, the increased use of ultrasonic imaging in obstetrics is not so important as its use in abdominal imaging. Endoscopic imaging of the gastrointestinal tract could have a significant impact on occupational exposure if it becomes as medically efficacious as barium contrast fluoroscopy. Magnetic resonance imaging has already had a significant impact on neuroimaging. Although this will impact significantly on patient exposure, it will have little influence on occupational exposure since computed tomography produces very little worker exposure. One area to watch is magnetic resonance imaging of flowing blood. It is already possible to obtain "magnetic resonance angiograms" of larger vessels. Since cardiac catheterization and special vascular procedures contribute greatly to medical occupational exposure, any non-ionizing procedures that could replace them could have a dramatic impact on future patterns of occupational exposure.

The past has seen great improvements in the sensitivity of x ray image receptors. This trend probably will not continue. Photon absorption characteristics of most image receptors are within a factor of two of maximum, and most conventional x ray imaging is performed today under noise-limited conditions. Under these conditions, sensitivity can be improved only at the expense of image quality. Developments in digital image receptors could offer some increased sensitivity due to improved rejection of scattered radiation; however, the cost effectiveness of current technology remains highly questionable. Developments in new imaging technologies are driven primarily by medical efficacy and cost effectiveness. Patient dose is a concern only insofar as image quality is maintained, and occupational exposure is often considered only as an ancillary issue.

The discipline of medical radiation protection has little control over the "source term", that is patient exposure, which drives occupational exposure. The only formal control exercised over medical occupational exposure is that which influences the fraction of patient exposure reaching the worker. Traditional approaches to control are "time, distance, shielding and training" as they are organized into a formal program. Future trends are likely to involve the administration of radiation protection more than technological developments. Improvements in individual and collective occupational doses will come about primarily through ALARA programs, since we have already demonstrated that dose limits are almost universally met.

Successful implementation of ALARA programs will require a greater understanding of medical occupational exposure and a greater attention to radiation protection than is evidenced in many medical institutions. These changes will be coincident with the fragmentation of medical imaging and greater efforts to contain medical costs. Outpatient imaging centers that are trying to perform barium enemas in competition with the center next door will have to be convinced to spend resources on radiation protection activities that yield no direct financial return. This effort will more than likely occur in an environment less driven by regulatory concerns. No institution analogous to the Joint Commission for Accreditation of Healthcare Organizations (JCAHO) exists for free standing private clinics.

Risk Assessment and Monitor Placement

All of the dose equivalent data discussed above are presented with one major caveat. They are derived primarily from commercial monitoring services with no adjustment for the special nature of medical occupational exposure. The recent study by the EPA assumed that monitor readings represent whole body dose equivalent (1). The NCRP used the same EPA data in Report 93 (8). It is well known that this assumption is not very accurate. Most medical occupational exposure arises from fluoroscopy, and almost all workers are shielded by lead aprons during these procedures. Yet the monitor is often worn outside of the apron and yields a gross overestimate of wholebody dose equivalent. Furthermore, the radiation environment is non-uniform, and the penetrability of the x ray can vary greatly from case to case. Such circumstances can make the monitor reading very unrepresentative of whole body dose equivalent.

The issue of "where to wear the film badge" has been debated for several years (11, 12, 13). The NCRP has given conflicting guidance, with Reports 48 and 59 (14, 15) recommending that monitors be worn outside protective apparel and Report 57 (16) recommending that a single monitor be worn underneath protective apparel. Reports 57 and 68 also advised that it may be appropriate to wear two monitors (16, 17). The ICRP has taken a similarly ambivalent position, in one case recommending that a single monitor be placed in a position representative of the most highly exposed surface (18), and in another case recommending that more than one dosimeter may be desirable if a lead apron is worn (19). Both organizations suggest that a qualified expert should be consulted for an appropriate solution to the issue. In many cases the location of a single monitor is dictated by state law.

Fundamental issues are at stake on this dilemma, and its solution requires examination of the premises and intentions of radiation protection. As reviewed by Meinhold (20), the initial objectives of radiation monitoring were based on the demonstration of adequate safety as defined by maximum permissible doses (MPD) to critical organs. The emphasis on finding the location that yields the highest fraction of the MPD to an organ is entirely consistent with the MPD approach. The monitor reading is used to demonstrate achievement of a particular goal, namely that a critical organ dose has not been exceeded. The question of whole body dose equivalent does not arise because adequate whole body protection is assumed to follow from the protection of critical organs. The MPD approach to protection also places fewer demands on the accuracy of personnel monitoring. It was recommended that for readings approaching the MPD, a measurement accuracy of ± 30 percent should be achieved while at the level of 1/4 of the MPD, an accuracy of ± 100 percent was acceptable (16). Approximately 95% of medical workers fall into the later category of readings.

Recently both the ICRP and NCRP have adopted a risk based approach to radiation protection. The risk of a radiation detriment is assumed to be linearly proportional to the effective whole body dose equivalent as calculated from the sum of weighted organ doses. One advantage of this approach is that it permits quantitation of risk and comparison to other occupations that do not involve radiation. It also facilitates a quantitative approach to the optimization of radiation protection. Even within the context of risk-based protection, the ICRP has recognzied that a large uncertainty in whole body dose equivalent is acceptable when doses of less than 10 mSv (1000 mrem) are considered.

Monitoring records have also been accessed with increasing frequency for epidemiological studies and legal proceedings. Both of these uses attempt to impute risk from exposure, either through statistical inference or as a matter of probable cause. In these applications, monitoring records may be used in an uncritical manner, incuding ascribing greater accuracy than warranted to low level exposure records. This presumption of accuracy builds pressure to improve the accuracy of personnel monitoring.

Recent developments in risk based protection, together with scientific and legal requirements are placing personnel monitoring into a new situation. The direction of change seems to be toward a requirement for greater accuracy, in terms of both magnitude of dose and its spatial distribution. The issue is fraught with difficulties that must be examined in considerable detail before official recommendations can be made. It is evident that this issue should be considered by the NCRP.

The adoption of a risk based system of radiation protection creates new problems for personal monitoring. It is necessary to know the effective whole body dose equivalent and implementation of risk based radiation protection requires that one be able to monitor or derive doses to all organs. This is no easy task in the case of medical occupational exposure. Some circumstances, such as those that occur in nuclear medicine or radiation therapy, involve high energy, reasonably isotropic sources of radiation that may accurately be assessed regardless of the location of the monitor. Diagnostic radiology workers will present the greatest challenge. A single monitor may be used to derive doses to all organs if sufficient information is available concerning the energy and spatial distribution of radiation. However, the conversion factor from monitor reading to whole body effective dose equivalent can vary by a factor of four depending on x ray kilovoltage and location of the source. Some departments have several different thickness lead aprons available, necessitating a different conversion factor for each thickness. Hence, the use of a single monitor will require much more accurate knowledge of the circumstances of each exposure. At minimum this will require greater record keeping and may be a practical impossibility. Use of multiple monitors may not offer much help. To be meaningful each monitor must be worn in the same location each day. Since the collar and under apron monitor readings may differ by a factor of ten, an interchange of monitors on only one day per month could result in an error of up to 30 percent.

Conclusion

Medical occupational radiation protection has been very successful over the past twenty years. Very few workers approach dose limits, even without taking into account effect of shielding aprons, and collective dose has decreased even while utilization has increased. Past improvements were probably due in large part to improvements in image receptor technology. Since patient exposure, the source term of medical occupational exposure, is outside the control of radiation protection, future improvements are likely to lie in the application of the ALARA principle to the medical work environment. This will require the radiation protection community to embrace the risk based system of dose assessment and to solve the problem of accurate monitoring of effective whole body dose equivalent.

References

1. EPA. Environmental Protection Agency. *Occupational Exposure to Ionizing Radiation in the United States: A Comprehensive Review for the Year 1980 and a Summary of Trends for the Years 1960–1985* by Kumazawa, S., Nelson, D.R. and Richardson, A.C. EPA 520/1-84-005 (National Technical Information Service, Springfield, Virginia).

2. ACR. American College of Radiology. "Survey of diagnostic imaging scientists and engineers," (American College of Radiology, Reston, Virginia) (1988).
3. Radecki, S.E. "Diagnostic radiology usage in ambulatory and hospital care," Radiology, **167,** 857–860 (1988).
4. Bunge, R.E. and Herman, C.L. "Usage of diagnostic imaging procedures: A nationwide hospital study," Radiology, **163,** 569–573 (1987).
5. Mettler, F.A. "Diagnostic radiology: usage and trends in the United States, 1960–1980," Radiology, **162,** 263–266 (1987).
6. Johnson, J.L. and Abernathy, D.L "Diagnostic imaging procedure volume in the United States," Radiology, **146,** 851–853 (1983).
7. Johnson, D.W. and Goetz, W.A. "Patient exposure trends in medical and dental radiography," Health Phys., **50,** 107–116 (1986).
8. NCRP. National Council on Radiation Protection and Measurements. *Ionizing Radiation Exposure of the Population of the United States* NCRP Report No. 93 (National Council on Radiation Protection and Measurements, Bethesda, Maryland) (1987).
9. NCRP. National Council on Radiation Protection and Measurements. *Quality Assurance for Diagnostic Imaging Equipment,* NCRP Report No. 99 (National Council on Radiation Protection and Measurements, Bethesda, Maryland) (1988).
10. Weinstein, D. "Imaging studies down overall, but new modalities show increase," Radiology Management, **8,** 15–17 (1986).
11. Wiatrowski, W.A. "The recommended location for medical radiation workers to wear personnel monitoring devices," (Letter) Health Phys., **38,** 433 (1980).
12. Bushong, S. (Letter) Health Phys. **40,** 258–259 (1981).
13. Bushong, S. (Letter) Health Phys. **42,** 242–244 (1981).
14. NCRP. National Council on Radiation Protection and Measurements. *Radiation Protection for Medical and Allied Health Personnel,* NCRP Report No. 48 (National Council on Radiation Protection and Measurements, Bethesda, Maryland) (1976).
15. NCRP. National Council on Radiation Protection and Measurements. *Operational Radiation Safety Program,* NCRP Report No. 59 (National Council on Radiation Protection and Measurements, Bethesda, Maryland) (1978).
16. NCRP. National Council on Radiation Protection and Measurements. *Instrumentation and Monitoring Methods for Radiation Protection,* NCRP Report No. 57 (National Council on Radiation Protection and Measurements, Bethesda, Maryland) (1978).
17. NCRP. National Council on Radiation Protection and Measurements. *Radiation Protection in Pediatric Radiology,* NCRP Report No. 68 (National Council on Radiation Protection and Measurements, Bethesda, Maryland) (1981).
18. ICRP. International Commission on Radiological Protection. *General Principles of Monitoring for Radiation Protection of Workers,* ICRP Publication 35 (Pergamon Press, New York) (1982).

19. ICRP. International Commission on Radiological Protection. *Protection Against Ionizing Radiation from External Sources Used in Medicine*, ICRP Publication 33 (Pergamon Press, New York) (1980).
20. Meinhold, C.B. "The impact of the probability of causation on the radiation protection program," Health Phys. **55,** 357–377 (1988).

Discussion

JOHN CAMERON (University of Wisconsin): Bill, do you have any relative numbers for the average exposures of radiation workers in medical situations that compare to the exposures in nuclear power plants? How do they compare? Factor of 10 less on the average or how would you say?

WILLIAM HENDEE: The exposures are much less. I am not sure what the average quantitative number should be; it could be as much as 8 or 10 less. It is considerably less on the average. And the distribution is very heavily skewed to very low exposures in medicine.

JOHN CAMERON: Typically below, what, a hundred?

WILLIAM HENDEE: Well, the average is below 0.7 millisieverts per year and only a few percent ever exceed 1 millisievert per year. So the exposures are very low. And I think it presents an interesting question of how low is low enough before you begin to apply your resources to other areas of exposure where you could make a greater difference?

HYMER FRIEDELL (Case Western University): There are a couple of points. First of all I enjoyed your dissertation and your excellent review of problems that exist. There are a couple of points that I wondered whether you might wish to address. One is the legal question and this has some interesting aspects. Obviously there must be uncertainty because if we never made any errors, there would be no concern. Obviously, errors exist. Once there is uncertainty the only way you minimize uncertainty in the present is by redundancy. Therefore, taking a page from information theory, more studies of exactly the same nature are better and reduce the uncertainty. The saving feature is some studies that were made a long time ago showing that, even though you degrade the image considerably, the accuracy doesn't go down proportionately. Therefore, some decision has to be made about risk rate. We get back to the same question, how much reliability do we wish? And how safe do we want to be in these various aspects?

WILLIAM HENDEE: Just a very brief response because I think you have been eloquent in explaining the problem. Should you decide not to take an x ray exam, and it turns out that the patient suffers an adverse consequence, now it is not necessarily true that had you taken the x ray exam you could have prevented the consequence. Still it may be that a suit is filed against you simply because you didn't take the x ray exam. That's the end in itself. Physicians in the health care setting are very sensitive to this problem and are caught in a terrible trap. In many cases they have really little choice often except to exercise the redundancy that you point out because the non-exercise of that redundancy may in itself be grounds for legal action.

ALAN BRODSKY: Thank you for your interesting and informative talk. I found myself wondering whether you could say a word about how well we know the uniformity in collection from year to year of these exposure data. Are our exposure data collection systems at a quality assured and adequate?

WILLIAM HENDEE: It's hard for me to answer that question Alan, because we use the data from groups like EPA and other groups and yet we are not intimately involved in the collection process. My guess is that this is a troublesome area because of the diffusion of the medical radiation work force into every nook and cranny across our society. Most of these people employ x rays, which means that they're under the supervision, in general, of the State Health Department but they are not under the supervision of the Nuclear Regulatory Commission. So that one might wonder just how adequately we really can sample and get accurate measures of what the radiation exposures are. When we get those measures we know they are film badge readings. We don't know where the film badge was. We recognize that if we were to take those film badge readings and convert them into a meaningful whole body dose equivalent, then the doses would change substantially.

ALAN BRODKSY: I see there's still controversy in the recent Health Physics Journal in letters to the editor on this. Maybe the NCRP could help resolve that.

WILLIAM HENDEE: Yes.

JOEL GRAY (Mayo Clinic): Bill, I enjoyed your presentation. There was one area you mentioned-poorly trained technologists and I know we do have a significant shortage of techs right now. I have an area of concern, and I think it's even greater than that and that's the number of x ray examinations being carried out by non-technologists in non-radiology offices. And we in the state of Minnesota have seen this and refer to them as the secretaries, nurses, and janitors that are taking the x rays.

WILLIAM HENDEE: Well, Joel, I think you are likely to see more of that in the future. We really cannot get technologists to come into training and I understand why they don't want to. There's very little vertical mobility, the salaries are poor and they have nowhere to aspire to once they enter the field. We need to solve those problems internally within the field of medicine. But I think that unless we begin to solve them fairly quickly we're going to see a lot of people that are relatively poorly trained for three reasons. One is the reason you mentioned, that people are going to be forced to utilize non-trained people in their offices. The second is that the quality of the applicants continues to decline, so we have less satisfactory material to begin with in our training programs. And the third reason is that I'm concerned about the quality of some of our training programs in the sense that we have migrated out of hospitals and into community colleges and into the junior colleges for conduct of these training programs. I'm not sure that we have an adequate mechanism to insure that these people really get trained in the day by day routine of what it means to be a radiologic technologist and to do radiologic technology. We may have some people who have quite a bit of course work but perhaps less experience than we would really like to see them have.

CHARLIE MEINHOLD: Just a quick one from me, Bill, before you step down. I was intrigued with Jack Schull's presentation this morning and thinking about the large proportion of workers in medicine that are young women. And I wondered if you had a reaction to that question with regard to the occupational exposure. Will more attention to the potentially pregnant female cause us difficulties in the medical field?

WILLIAM HENDEE: The answer is yes, I think we've got to take these data into more cognizance than we have in the past and it's not only the young females who are working in radiology but it's also the young female patients. I think we've got to be more concerned about that. We've been concerned about it but perhaps not to the degree that we should have been.

Reconstructing Historical Exposures to the Public from Environmental Sources

John E. Till
Radiological Assessments Corporation
Neeses, SC

Abstract

Considerable attention has recently been given to estimating exposures to individuals who lived near major sources of radionuclides released to the environment during atmospheric nuclear weapons testing and early operation of nuclear weapons production facilities. Most of these releases occurred many years ago and therefore make determinations of dose to specific individuals complex and costly. Dosimetry data coupled with epidemiology adds to our data base for disease induction attributable to radiation exposure. As a direct result of this historical analysis, the state-of-the-art of radiological assessment has advanced significantly not only in the models being applied, but also in techniques used for data collection and estimation of uncertainties. This article discusses what is being learned from dose reconstruction and explains how analysis of the past is leading to new methodologies for addressing exposures to the public in the future.

Introduction

In keeping with our theme on "Radiation Protection Experience" I would like to address one aspect of exposures to the public that is playing a major role in advancing the science of dose assessment and radioepidemiological reaseach, "The Reconstruction of Historical Exposures to the Public from Environmental Sources." I would like at the outset to give credit to the National Cancer Institute, The University of Utah, and to the Department of Energy which have supported my efforts in this research.

As a prelude, however, I would like to briefly review the major historical milestones that have brought us to this current vantage point of radiological assessment. It has often been said that we know more about radionuclides in the environment than any other pollutant known to man. Although the validity of this statement is subject to challenge, it is apparent that there was an early concern for radiation protection when nuclear programs were developing at U.S. Government and University Laboratories.

Beginning in the 1950's, assessment and measurement of fallout from atmospheric nuclear weapons testing was carried out on a large scale throughout the United States. These data are now a fundamental resource for the comprehensive dose reconstruction studies ongoing today.

Initiation of the Plowshare Program in 1957 created an immediate need for predicting the dispersion and ultimate fate of radionuclides that might be vented to the atmosphere or enter ground water and expose man. By the late 1960's the commercial nuclear power program began to gain momentum and methodologies for assessing public exposures were applied in studies addressing impacts from operations of these facilities.

The National Environmental Policy Act, NEPA, which was signed into law in 1970 required Federal Agencies to prepare comprehensive assessments of all potential environmental impacts resulting from major projects under their charge. The U.S. Nuclear Regulatory Commission's Regulatory Guide 1.109 (1) recommended specific models and parameters to be applied in assessment of public exposures.

Most dose assessments that antedate NEPA were prepared using conservative methodologies. However, NEPA gave birth to public hearings, reviews by other government agencies, and special interest groups. There was pressure to reduce the calculated exposures further and thus the concept of "As Low As Reasonably Achievable," ALARA, was derived.

The publication of WASH-1400 (2) was the first comprehensive application of probabilistic risk analysis to addressing public exposures. In

1984, the NCRP published Report No. 76 (3) on Radiological Assessment, which looked at the state-of-the-art of assessing public exposures and emphasized the needed direction for the future, including screening models and uncertainty analysis.

Finally, the dose reconstruction studies addressed in this article are playing a key role in improving estimates of exposures to the public from environmental sources. Perhaps the most important aspect of the dose reconstruction efforts is the emphasis on deriving a "best estimate" of dose rather an upper bound (or conservative) value which has often been the objective of past radiological assessments. The discussion which follows is an overview of how dose reconstruction is being used to advance the science of radiological assessment.

Ongoing Dose Reconstruction Studies

Our investigation of historical exposures, or dose reconstruction, has been the major source for support of methodology development and model validation for the past decade. These studies have dealt with estimating exposures to the public from releases of radionuclides from nuclear research and production facilities, the accidents at Three Mile Island and Chernobyl, and the atmospheric testing of nuclear weapons at the Nevada Test Site.

There are four major dose reconstruction projects currently ongoing. The Offsite Radiation Exposure Review Project (ORERP), supported by the U.S. Department of Energy was initiated in 1979 to estimate exposures to persons living in the proximity of and downwind from the Nevada Test Site. This study was the earliest of these four and set out many of the fundamental processes which have been applied in the projects that followed (4).

The National Cancer Institute Radioiodine Project was directed by Congress and has the objective to "Develop valid and credible assessments of exposure to ^{131}I that the American people received from the Nevada atmospheric bomb tests." Therefore this study deals with exposure to the entire population of the continental U.S. and will provide estimates of absorbed dose to thyroid for the centroid of approximately 3,000 counties in the U.S. (5).

The Hanford Environmental Dose Reconstruction Project is estimating doses resulting from releases from the Hanford Reservation during the early years of operation of the Hanford Site. This study is considering a different source term that originated from the production of special nuclear materials for nuclear weapons rather than weapon detonation. This research is the newest of the four.

The University of Utah Fallout Study includes a dose reconstruction and epidemiological analysis that focuses on individuals who lived primarily in Utah during the atmospheric nuclear weapons testing period. This project will be used in this article as an example of how dose reconstruction is making a significant contribution to our ability to assess exposures to the public from environmental sources.

The University of Utah Fallout Study

The Utah Study has two parts, a leukemia case-control study and a thyroid cohort study. The objectives of the leukemia study are to identify and verify independently all cases of leukemia in residents of Utah diagnosed between 1950 and 1980, to identify appropriate controls from the population of the State of Utah, to estimate individual bone marrow dose for cases and controls, and to conduct appropriate statistical analyses to detect excess risk as a function of external gamma dose arising from fallout.

Leukemia case-control study

In order to be included as a case or control, study subjects must have been a resident of Utah at the time of death, must have had a Utah death certificate, must have been born prior to November 1, 1958 and died between January 1, 1952 and December 31, 1981, and must have been a member of the Church of Jesus Christ of Latter Day Saints (LDS). The cases consisted of all deaths from leukemias meeting these criteria. The controls consisted of a matched random sample of all other deaths. Membership in the LDS Church was a criterion since the Church has extensive records on member residences that permitted a reconstruction of the individual's exposure which is highly dependent on residence location.

The dose to bone marrow is estimated with the following expression.

$$D = \sum_{i=1}^{n} \sum_{j=1}^{m} X_{ij} \times F_{ij} \times DCF \times MF$$

where,

D = total absorbed dose (gray) to active bone marrow for a specific case or control received from all events at all locations,

X_{ij} = infinite-time integrated external exposure in air at location 'i' due to weapons fallout from event 'j',

n = number of locations at which subject resided and which received fallout.

m = number of fallout events which deposited fallout at location 'i',

F_{ij} = fraction of infinite-time exposure at location 'i' from event 'j' received during the time spent at location 'i',

DCF = age dependent dose conversion factor (gray per ergs per g of air), and

MF = modifying factor to account for building shielding which reduces the absorbed dose.

Keys to establishing the best estimate of dose in this equation are knowing the exposure at the location of interest and knowing where the individual lived and the length of time at that residence. Two critical data sets helped us with this information and significantly reduced the uncertainty of the dose estimates in the leukemia study. The deposition data bases, known as the Town and Country Data Base (6, 7), were developed by the Offsite Radiation Exposure Review Project and provided the fundamental estimates of deposition that helped determine the amount of fallout at each residence location.

The second critical set included the deceased member file and the census records of the LDS Church, which helped to establish a detailed account of each subject's location and period of residence. Figure 1 illustrates the key elements of the dose reconstruction for the leukemia study. The subject is identified as an eligible case or control, the residence history is determined using the church census records or the deceased member file. (Telephone and city directories were used to verify the residence histories and to help fill in gaps missing in the data base.) Total exposure was estimated using the deposition data bases. The dose conversion factor was applied to obtain the dose assignment.

An example dose reconstruction is shown in Table 1 for a study subject. This subject lived in Salt Lake City from 1950 to October 1952 then moved to Indiana and New York until December 1956 when he returned to Salt Lake City until 1964. Important fallout events contributing to exposure are listed in the left column, the dates of the shot in the center, and exposure in mGy in the right column. For this subject, most of the exposure came from shot EASY, which contributed 64% or 1140 mGy of the total 1775 mGy dose.

This slide does not show the age of the subject but the dose conversion factor used at each stage of the calculation would be a function of age. A generic factor for shielding of 0.5 was applied to all subjects.

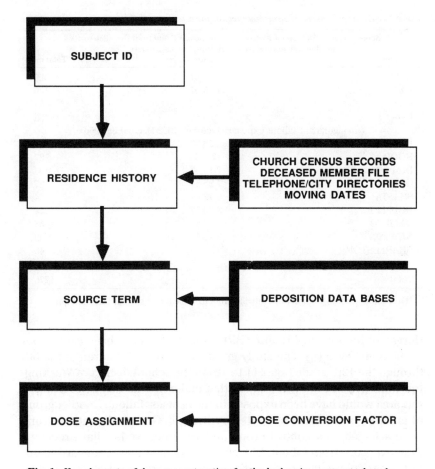

Fig. 1. Key elements of dose reconstruction for the leukemia case-control study.

Altogether, absorbed dose to bone marrow was estimated for 1177 cases and 5330 controls. Although the results of the dose reconstruction in the leukemia case-control study are still preliminary, current best estimates of dose to the active bone marrow range between 0 and 30 mGy for cases and controls. To date, the results of the epidemiological analysis have not yet been completed.

Thyroid cohort study

The first objective of the thyroid cohort study was to relocate and reexamine the thyroids of subjects who had participated in the original

TABLE 1.—*Example dosimetry reconstruction for a leukemia case-control study subject*

Subject lived in Salt Lake City from 1950 to Oct 1952 and from Dec 1956 to 1964
(lived in Indiana and New York from Oct 1952 to Dec 1956)

Fallout event	Date	Total mGy*
ABLE	1-27-51	26
EASY	11-5-51	1140
FOX	5-25-52	51
HOW	6-5-52	33
(no significant fallout exposure October, 1952-December, 1956)		
BOLTZMANN	5-28-57	36
HOOD	7-5-57	27
DIABLO	7-15-57	103
OWENS	7-25-57	27
SHASTA	8-18-57	74
DOPPLER	8-23-57	124
GALILEO	9-2-57	58
NEWTON	9-16-57	33
CHARLESTON		43
Total		1775

*Only shots making significant contributions to dose are shown in this example.

Bureau of Radiological Health (BRH) study which had been carried out in the mid 1960's (8). The study group consisted of children in the 5th through the 12th grade (ages 11 to 18) in the school district of Washington County Utah. It was assumed that children who had been living in this area would have been exposed to the heaviest fallout. Another group of children in the same age category living in Graham County, Arizona were selected as a control group and were believed to have received little exposure from fallout. The clinical examinations were carried out by teams of physicians or practitioners who had received special training for the project.

The second objective of the thyroid cohort study is to estimate absorbed dose to the thyroid for the study subject who could be located and reexamined. This task was carried out with the use of surveys of the parents of the cohort to determine dietary and lifestyle habits of the subjects and through a careful reconstruction of dairy farming and milk distribution practices ongoing at the time of the exposure.

The final objective was to conduct an epidemiological analysis incorporating the results of the clinical examinations and the dose reconstruction.

There were approximately 4819 subjects in the original cohort. Approximately 4200 of the original group were relocated and reexaminations were carried out on 3122 of those or about 65%.

Absorbed dose to the thyroid is the sum of the dose from ^{131}I and ^{133}I for each shot. It is determined with the calculation shown below.

$$D = \sum_{n=1}^{\#shots} \sum^{131+133} \{[\frac{Cm}{dep} * dep] * CR * (e^{-\lambda t}d) * DCF\}$$

where,

D	= absorbed dose to thyroid for cohort (Gy),
$\frac{C_m}{dep}$	= integrated radioiodine concentration in milk per unit deposition (Bq-d/L per Bq/m^2),
CR	= consumption rate of milk (L/d),
$(e^{-1}{}^td)$	= decrease in concentration due to delay between production and consumption, and
DCF	= dose conversion factor (Gy/Bq).

In the thyroid cohort study, the deposition data bases of the ORERP study were again used to determine the amount of fallout present at a given location. However, in order to derive a best estimate of dose, critical data were needed on dairy practices and milk distribution as well as information on diet and sources of milk from study subjects. To get this information extensive surveys were conducted on dairy producers and processors in the region and on parents of the study cohort. These data were entered into the dose calculation for each subject in order to obtain the best estimate of dose possible. Further, the data from the surveys were used to develop a set of default values that were applied in the study when person-specific data could not be obtained. The content and conduct of these surveys proved to be a significant contribution to this work and will be of great benefit to future dose reconstruction efforts.

Figure 2 illustrates the key elements of the thyroid cohort study. Subjects were identified as being a member of the original BRH cohort. Their residences, diet, lifestyle and medical history were determined from a telephone survey of parents (or closest living relative). Although questions in the cohort survey contained detailed information about diet, consumption rates, and sources of food products, the most critical information needed was whether or not an individual consumed milk and was it produced locally. It is believed that most respondents could recall this fact with a high degree of certainty and knowledge of fresh milk consumption is the most critical element of the dose calculation.

A second survey was carried out to identify milk processors, producers and specific feeding regimes used by the farmers of the area. These data were essential in reconstructing an accurate account of the path of

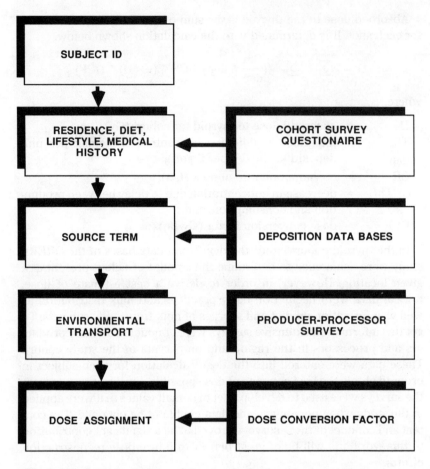

Fig. 2. Key elements of dose reconstruction for the thyroid cohort study.

radioiodine through the food chain to the cohort. Figure 3 illustrates the complexity of a typical feeding regime for a single farmer. The amount of radioiodine ingested by dairy animals depends significantly on the amount of time a cow spent on pasture and the fraction of the diet on a dry matter basis that the pasture contributed. As one follows the course of the year, it is shown that the diet consisted of silage and old hay which was essentially void of any radioiodine until about May 1 when pasture began. At that time there would have been considerable potential for contamination of milk. There were approximately 800 distinct feeding regimes ultimately included in the data base of study and applied in the calculation of a subject's dose, depending on the source of milk identified in the survey.

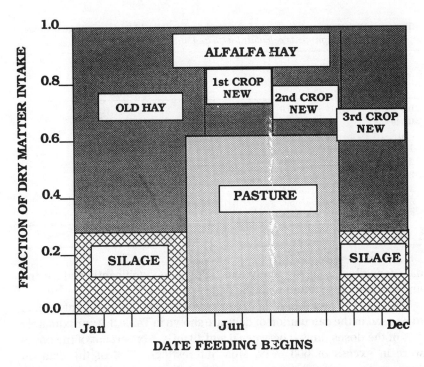

Fig. 3. Example of feeding regimes developed for calculating the environmental transport of radioiodine in the thyroid cohort study. Approximately 800 unique feeding regimes were ultimately prepared based on results of the producer-processor survey.

Table 2 illustrates a typical dose reconstruction for a study subject. This individual received a total dose to thyroid of 380 mGy from ^{131}I and 14 mGy from ^{133}I. The majority of the dose came from shot HARRY. As expected, dose was highly dependent on the time that a particular shot occurred and the feeding regime applied at the milk source, pointing out the high importance of knowing specific residence history and dietary

TABLE 2—*Example thyroid dosimetry reconstruction for study project*

Test name	Absorbed dose to thyroid (mGy)	
	^{131}I	^{133}I
HARRY	330	13
PRISCILLA	0.3	0.01
SIMON	9.5	0.29
SMOKY	26	0.4
TESLA	6.5	0.07
ZUCCHINI	7.5	0.13
Total	380	14

TABLE 3—*Preliminary estimates of absorbed dose to thyroid and percentage of dose assignments calculated with and without default data*

Range of dose mGy	mGy	Defaults used No (% of total)	Yes
0.0	0.0	8	1
>0.049	32	11	
50	99	12	4
100	199	9	2
200	299	5	1
300	399	2	1
400	499	2	1
500	599	2	1
>600		5	1
Total		77	23

data for subjects in the study and helping to significantly reduce the uncertainties associated with the dose estimates.

The range of doses and the frequency with which default values had to be used in the calculation of dose are shown in Table 3. Approximately 43% of the doses ranged between 0 and 49 mGy. Six percent of the doses were in excess of 600 mGy. Most interestingly, 77% of the data on consumption and lifestyle were supplied directly by the subjects and default values (information inserted when a respondent could not recall specific values) had to be applied in only 23% of the subject doses. It is emphasized that these results are still considered preliminary.

Conclusions

The University of Utah Fallout Project is an example of how doses to the public are being reconstructed from historical releases of radionuclides long after the exposures occurred. This research has yielded several noteworthy contributions to the advancement of radiological assessment.

Specifically, dose reconstruction has provided the major source of support for radiological asessment research over the past decade. Without the resources invested in these studies, our ability to predict the environmental transport and fate of radionuclides released to the environment would have been significantly reduced.

The dose reconstruction studies have resulted in significant advances in the state-of-the-art of dose assessment with particular achievements in uncertainty analysis, the design and implementation of surveys for dose assessment, and our capability to provide a best estimate of dose rather than an upper bound value.

Perhaps the most important contribution of the dose reconstruction studies is the potential to apply these techniques to assessment of nonradioactive substances in the future.

References

1. NRC. U.S. Nuclear Regulatory Commission. "Calculation of Annual Doses to Man from Routine Releases of Reactor Effluents for the Purpose of Evaluating Compliance with 10 CFR Part 50, Appendix I," Regulatory Guide 1.109 (U.S. Nuclear Regulatory Commission, Washington) (1977).
2. NRC. U.S. Nuclear Regulatory Commission. "Reactor Safety Study," WASH 1400 (U.S. Nuclear Regulatory Commission, Washington) (1975).
3. NCRP. National Council on Radiation Protection and Measurements. *Radiological Assessment: Predicting the Transport, Bioaccumulation, and Uptake by Man of Radionuclides Released to the Environment*, NCRP Report No. 76 (National Council on Radiation Protection and Measurements, Bethesda, Maryland) (1984).
4. Church, B.W., Wheller, D.L., Campbell, C.M. and Nutley, R.V. "Overview of the Department of Energy's off-site radiation exposure review project (ORERP)," Health Phys. (in press).
5. Wachholz, B.W. "Overview of the National Cancer Institute's activities related to exposure of the public to fallout from the Nevada test site," Health Phys. (in press).
6. Thompson, C.B. and Hutchinson, S.W. "The town data base: estimates of fallout times and exposure rates near the Nevada test site (NTS)," U.S. DOE Report NVO-322 (U.S. Department of Energy, Las Vegas, Nevada) (1988).
7. Anspaugh, L.R. and Beck, H.L. *The County Data Base: Estimates of Exposure Rates and Times of Arrival in the Off-Site Radiation Exposure Review Project (ORERP) Phase II Area*, U.S. Department of Energy Report NVO-320 (U.S. Department of Energy, Las Vegas, Nevada) (1988).
8. Rallison, M.L., Dobyns, B.M., Keating, F.R., Rall, J.E. and Tyler, F.H. "Thyroid disease in children: a survey of subjects potentially exposed to fallout radiation," Amer. J. of Med., **56,** 457-463 (1974).

Discussion

WILLIAM TEMPLETON (PNL): John, you dealt a lot with the Utah study, but you are involved with them all. What are the major differences you see?

JOHN TILL: Well, in particular, there are several key differences between the Hanford study and the O-Rep, the Utah, and the NCI radiodine project. Of course they all have many commonalities. Obviously, as I've already mentioned in the Hanford study, we have a different source term. We're not dealing with atmospheric weapons testing, we're dealing with a spectrum of radionuclides from a production facility. I think though I'd like to mention one thing about the Hanford study that makes it stand out. It's not in the technical manner in which the work is being carried out, but it's how the project is being managed. For those of you who don't know the Hanford study, which is still new, the technical work is actually being done by Battelle. It is being directed by an independent technical steering panel; completely separate from Battelle, with it's own authority and it's own independence. And the major element of that is that the public is involved with this technical steering panel and we're trying to involve the public all along the way, so that hopefully when we're finished, we will have had their input throughout the process and that what we do will be credible.

AL TSCHAECHE (Westinghouse, Idaho Nuclear Corp.): You gave us some numbers; some of our friends in the EPA calculate doses down to 10^{-75} rem. Do you have a lower limit of dose beyond which you will not bother to make an assessment?

JOHN TILL: We don't yet, but we will. In fact, that is a key item on our agenda in the Hanford study. As far as the other studies are concerned, there is no limit.

ED WRENN (University of Utah): John, I just want to point out for the audience that there were three studies conducted at the University of Utah on fallout. I directed the other two and they were measurement studies designed to see whether we could measure today things in the environment that would tell us about fallout from the past. The study of external doses that Ed Haskell and I developed by measuring thermoluminescence of bricks basically verified that the fallout measurements made in the fifties with the survey meters among other things gave reasonable values as re-measured today. And the other study deals with measurements of iodine-129 in thyroids obtained at autopsies from the forties and fifties. And plutonium in bones of people who had been walking around then but died in the eighties. And I for one think it's very important in reconstructing doses to have measurements as well as models so that one can verify that the models indeed predict something reasonable.

JOHN TILL: That's a very good point Ed, thank you.

MIKE RYAN (CHEM - Nuclear Systems): John, you mention how you verified location of individuals who are accepted either into the control group or the cohort group. For example, dietary habits and other things that are less trackable and less easy to verify. How did you deal with uncertainies in those kinds of data?

JOHN TILL: Well we really didn't, I mean we had to assume that when an individual told us they drank milk or did not drink milk that they knew what they were telling us. On the other hand, we have included a factor of uncertainty in terms of the range of the amount of milk that was consumed. Unfortunately I didn't have the time to discuss or to share with you the uncertainty analysis on this work, which I believe is really pushing the science a new step forward. So it does include uncertainty, Mike, and in a very deliberate sense.

SYDNEY PORTER (Porter Consultants, Inc.): John, could you share with us what you feel the most significant advances are that you've made in your uncertainty analyses for this one study you've been talking about on the radioiodine work?

JOHN TILL: Well, I think that has to be lumped into just one statement about uncertainty, and it's the fact that we have tried to account for as many of the different processes as possible in the uncertainty analysis. It is a very quantitative analysis. It is not a qualitative uncertainty analysis. Every step along the way, whether or not a particular feeding regime was being applied on May 1st, we accounted for uncertainty plus or minus so many days. So just the whole process itself of uncertainty I think, was the greatest advance.

SYDNEY PORTER: Could you give us a rough estimate of what you feel the uncertainties were and what you've been able to knock them down to now? Can you give me some feel for this?

JOHN TILL: Well, it's pretty difficult to do that in one sentence without a lot of qualifiers. But on the leukemia study, and there are a lot of qualifiers for this, for uncertainties which we've been able to account for, we think we're dealing with about a factor of two. Now we've omitted certain things in this calculation and in this statement that I don't have time to go into. In the thyroid study we're probably talking about uncertainties on the order of a factor of five or so. And we feel pretty good about that. But again, there are a lot of qualifiers that go with that statement which I don't have time to discuss. Over all, a factor of two to five.

Historical Aspects of Medical Radiation Exposure and Protection

Fred A. Mettler, Jr.
University of New Mexico
Albuquerque, New Mexico

For purposes of this lecture an attempt has been made to assess exposures, absorbed dose, and frequency of diagnostic examinations. Population characteristics of those receiving examinations are presented and I will briefly trace the progress in radiation protection in relation to medical exposure over the last ninety years.

The period from 1895 to 1900 was, of course, marked by Roentgen's discovery of x ray sometime in early November 1895. Roentgen had been experimenting with a tube which emitted an unknown type of radiation. He had enclosed the tube completely in black cardboard and noticed that in spite of this he could see a shadow of his hand on another piece of cardboard painted with a fluorescent material.

The first medical x ray was performed on November 8, 1895 on the hand of Roentgen's wife, Bertha. Roentgen kept his findings a secret even from his laboratory assistants until he communicated the findings in the lst week of December, 1895. There was a fair dispute about who actually took the first x ray in the United States, but it is clear that by February of 1896 x rays of extremities and chest were being obtained in several U.S. cities. The actual exposure remains unknown although it is clear that exposure times ranged from several to many minutes. At least one chest x ray taken in 1896 at the Massachusetts Institute of Technology took 45 minutes (1).

There was extremely rapid technology transfer and during the course of 1896, Pupin had used intensifying screens which he obtained from

Edison (2). Manufacturing companies sprang up and in 1896 the Beale Manufacturing Company of St. Louis offered an x-ray apparatus for fifteen dollars. Although most exposures were made on glass plates until World War I, the use of film had actually been suggested in 1896. As one might expect with rapid development of technology, there was a need for transfer of information and the first annual meeting of the American Roentgen Ray Society was held in 1899.

During the decade 1901 to 1910 there was a large amount of attention paid to the design of x-ray tubes. Roentgen's original tube apparently operated on approximately 25,000 volts and the cathode ray impacted directly on the glass wall of the tube rather than on an anode. Unfortunately, gas filled tubes were difficult to control and the focal spot varied widely in size during the exposure. The typical x-ray tube did not change significantly in outward appearance and in general was suspended by wires and had little or no collimation. By 1902 fluoroscopy compression cones had been developed and by 1907 the first tilting table was in use for fluoroscopy.

This decade also saw wide use of radiation for therapeutic purposes. It was performed not only with relatively soft x rays, but also with radium, which had been discovered by Madame Curie in 1898. The hazards of radium uses were described as early as 1901 by Becquerel who carried some radium in his pocket and subsequently noticed a "burn".

From 1911 to 1920 there was much more development on x-ray tubes and in 1913 Coolidge developed a high vacuum, hot cathode tube, which contained a tungsten anode and cathode. Units of measurements were described in the first issue of the American Journal of Roentgenology in 1913 (3). The various exposure units were the Holzknecht unit which equalled approximately 2 Kienbock units which in turn was equivalent to tint b of Noires. The exact exposure that these units measured is unknown although it is mentioned in the article that three H units caused erythema and five H units were sufficient to cause temporary hair loss.

Throughout the first and second decades of the century it was a fairly common practice to utilize a bonnet fluoroscope. This is a device which was mounted directly on the head and had a sealed light type box about the eyes with a fluoroscent screen directly in front of the eyes. Under the circumstances the user would look directly into the beam and be able to visualize structures interposed between the tube and the intensifying screen, such as bones.

The last portion of the second decade brought World War I and forced significant changes in the design of equipment. Inability to obtain glass plates from Belgium caused a rapid change to film use. In addition there was a need to develop sturdy yet relatively affordable x ray generators and tubes for field use. Most of the x rays obtained during this time were utilized to locate either bullets or metallic fragments.

The decade from 1921 to 1930 saw the beginning of instrumentation for measurement of radiation and its use related to diagnostic exposure (4, 5). Although exact exposures and absorbed dose from medical exposures was difficult to measure there is some insight provided by Jerman in 1925 in an article (6) related to patient safety in diagnosis. Jerman indicated that when a 15 inch distance was used a total mAs should not exceed 600 "because of the danger of removing hair." Due to the danger of electrocution the same author also indicated that the high voltage circuit should be kept at least 20 inches from the patient. In 1924 Sampson gave an indication of the technique used for chest x rays (7), and indicated that 30 mA, 65 kVp, and between 1 and 4 seconds was usually sufficient. No tube-subject distance was given and therefore the exact exposure is difficult to ascertain but it certainly is in excess of 1 R.

An interesting side light is that technologies which we perceive as extremely modern and innovative today were often suggested and used many years ago. In 1925 Bartlett (8) suggested the use of teleradiology, and utilizing a seven minute time was able to transmit a hand radiograph from New York to Chicago. At that time the author suggested that teleradiology would be widely utilized throughout the United States within three or four years.

The decade from 1931 to 1940 was the beginning of organized radiation protection and its application to medical exposure. This came about primarily as a result of the development of continued concerns about radiologists safety (9), appropriate instrumentation and an effort to unify exposure units. These efforts were largely spearheaded by Taylor (10, 11).

In 1934 Hilt reported on the actual output on fluoroscopic equipment (12). He indicated that on at least several machines exposure rates ranged between 20 and 40 R per minute. One can see the benefit of improved instrumentation as well as regulation with present day instruments unable to legally exceed 10 R per minute, but usually not exceeding 5 R per minute.

In 1935 Turnbull (13), measured the exposure for several types of examinations. As an example, sinuses required an exposure of 15 R, kidney - 12 R, and Thorax - 0.4 R. Typical kilovoltage was 80. Present day exposures for chest radiographs are an order of magnitude less than those reported by Turnbull.

In 1936 Weyl, *et al.* (14) uncovered a problem which still remains with us to a large extent. He surveyed roentgenographic techniques for chest radiographs at 29 tuberculosis sanitoria and concluded that the quality of chest x rays was so different that no comparable diagnostic data could be exchanged. Of course, even at the present time it is known that there is often a wide distribution in exposures and absorbed doses when various facilities are surveyed. The characteristic distribution curve of exposures shows a long tail extending into the high dose regions.

Although there are many articles concerning reports of the uses of radium and x ray therapy for treatments of both malignant and benign disorders, the number of patients treated is unknown. An interesting article appeared in 1933 by Sayers on radium in medical use in the United States (15). This article, written by the Bureau of Mines, indicated that approximately 288 grams of radium had either been imported or produced in the United States and that surveys of medical institutions could account for only about 80 grams. The author estimated that radium was being utilized to treat 80,000 persons annually. The number of persons treated with x ray therapy was unknown.

The period from 1941 to 1950 was disturbed by World War II. Again the attention of x ray manufacturers was devoted towards production of sturdy and affordable equipment for battlefield conditions. One notable milestone occurred in 1941 when Lawrence (16) reported on the possibilities of utilizing radioactive material for diagnostic purposes and in fact obtained reasonable quality bone scans. Due to the war and the attempts at developing atomic weapons, all material from 1942 through 1946 remained classified. References to the use of radionuclides for medical diagnosis reappear in the literature in 1950 when Quimby and co-workers (17) described the utilization of iodine 131 in assessment of thyroid function.

The period from 1951 to 1960 saw the first significant analysis of the actual volume of practice in diagnostic radiology (18, 19). The early and mid-fifties also saw the rapid development of nuclear medicine with development of liver scans, brain scans, etc.

The two decades from 1960 to 1980 showed a fairly consistent increase in the number of radiologists, film usage, and machines. The number of procedures performed in the United States was about one-hundred million in 1964 and about two-hundred million in 1980. Data on age distribution of those receiving x ray examinations is available from both 1970 and 1980 and demonstrates as might be expected that over half the x rays are done on persons over the age of 45 (20).

Comparison of the types of examinations performed in 1950 and 1980 reveal some significant differences particularly with respect to the almost complete elimination of photofluorography. This practice, which was usually used in screening chest x rays, accounted for approximately one-third of all examinations done in 1950(18). The collective effective dose equivalent to the population for diagnostic x rays in 1980 has been estimated to be above 92,000 person-Sieverts.

During these two decades (1960 to 1980) significant changes in instrumentation and receptors significantly reduced the dose per examination. Examples of such improvement change from circular to rectangular collimators and finally to positive beam limitation collimators. These advances have been reported by Johnson and Goetz (21). This aspect alone may have well reduced the dosages to half of what they were previously. In the late 1970's and early 1980's the advent of rare earth screens as well as faster films also is a significant event leading to dose reduction factors of two to four (22). Also there was wide acceptance and use of gonadal shielding during this period resulting in markedly reduced gonadal doses.

The development of new technologies such as ultrasound in the late 1970's has lead to the virtual elimination of radiographs for pelvimetry measurements.

Significant dose reductions have been achieved through technological advances in mammography. The initial step was replacement of film mammography with xeromammography and subsequent utilization of intensifying screen systems. These changes as well as proper attention to compression and filtration probably have reduced the mammographic dose between 5 and 10 times from what was received in the late 1960's (23).

Nuclear medicine grew rapidly between 1960 and 1980 with reasonably good data concerning the numbers of examinations performed from 1972 on. In 1972 there were approximately three and a half million examinations performed, whereas in 1982 this had grown to about seven and a half million examinations annually. The age distribution of those receiving the examinations is also quite well known. In 1980 two-thirds of all nuclear medicine examinations were performed on persons over the age of 45. This is not surprising with the known applications for evaluation of metastatic and cardiac disease being the two prime indications for nuclear medicine studies. For 1980 the collective effective dose equivalent for nuclear medicine examinations is estimated to be about 30,000 person-Sievert.

Data on radiation therapy continues to be extremely difficult to obtain. Limited data suggests that in 1980 about 2,400 procedures utilizing tele and brachytherapy were performed per million population, and that about ten therapy machines existed in the U.S. per million population.

What has happened since 1980? Although some limited data is being obtained, the major studies carried out by the Federal Government in the sixties and seventies are no longer in progress due to the expense. Even if data is collected over the period of a year or two it needs time after that to be analyzed and finally published. At this point little actual numerical data on numbers of procedures is available from 1982 on. One can make some educated guesses however. The advent of magnetic resonance (MR) imaging will probably reduce the number of computerized tomographic (CT) examinations of the head and the spinal cord. This is important in terms of radiation protection since MR does not utilize ionizing radiation and CT is a relatively high dose procedure (often with surface doses in the range of three to five rad).

Even though medical exposure per examination has been reduced by one and possibly two orders of magnitude over the last ninety years, there are still areas for concern. Aggressive interventional techniques have been developed for removal of renal stones and for balloon dilatation of blood vessels. These are often very complex procedures and extended fluoroscopy times are commonly used. Cascade reported (24) that coronary angioplasty can result in exposure in the range of 100 R. These areas will require observation and monitoring in the future.

Without major advances in technology, the collective dose from medical uses of radiation will probably continue to increase. This will be the result of many factors including generalized aging of the population. Aging will result in presumably sicker patients requiring more diagnostic examinations and therapeutic procedures. Population size is also increasing, thus, unless there is reduction in utilization per person the total number of examinations will rise. The final major factor will be the increasing availability of various technologies, some of which contribute high absorbed doses. A typical example of this is the introduction of the CT scanner in the early to mid 1970's. The previously unique piece of relatively high dose equipment has now become commonplace in community hospitals.

References

1. Williams, F. "Reminiscences of a pioneer in roentgenology and radium therapy with reports of some recent observations," Am. J. Roentgenol. **13,** 253 (1925).
2. Martin, F. and Fuchs, A. "The historical evolution of roentgen-ray plates and films," Am. J. Roentgenol. **26,** 540 (1931).
3. Mackee, G. and Remer, J. "A technique for measuring the quality and quantity of the x ray," Am. J. Roentgenol. **1,** 49 (1913).
4. Fricke, H. and Beasley, I. "Measurement of the stray radiation in roentgen-ray clinics," Am. J. Roentgenol. **18,** 146 (1927).

5. Jacobsen, L. "Measurement of stray radiation in roentgen-ray clinics of New York City," Am. J. Roentgenol. **27,** 149 (1927).
6. Jerman, E. "X-ray technique: From the old to the new," Radiology **5,** 245 (1925).
7. Sampson, H. "Roentgenographic chest technique," Am. J. Roentgenol. **15,** 373 (1925).
8. Bartlett, G. "X-ray consultations of the future," Radiology **5,** 439 (1925).
9. Barclay, A. and Cox, S. "The radiation risks of the roentgenologist," Am. J. Roentgenol. **19,** 551 (1928).
10. Taylor, L. "International comparison of x-ray standards," Am. J. Roentgenol. **17,** 99 (1932).
11. Taylor, L. "Acurate measurement of small electric charges by a null method," Am. J. Roentgenol. **17,** 294 (1931).
12. Hilt, L. "The importance of measuring R output of roentgenographic equipment," Am. J. Roentgenol. **32,** 702 (1934).
13. Turnbull, A. and Leddy, E. "A method for determining the limits of safety in roentgenography," Am. J. Roentgenol. **34,** 258 (1935).
14. Weyl, C., Warren, S. and O'Neill, D. "A survey of chest roentgenographic technique," Am. J. Roentgenol. **35,** 526 (1936).
15. Sayers, R. "Radium in medical use in the United States," Radiology **20,** 305 (1933).
16. Lawrence, J.H. "The new nuclear physics and medicine, Caldwell lecture 1941," Am. J. Roentgenol. **48,** 283 (1942).
17. Werner, S., Goodwin, L., Quimby, E. and Schmidt, C. "Some results from the use of radioactive iodine in the diagnosis and treatment of toxic goiter," Am. J. Roentgenol. **63,** 889 (1950).
18. Donaldson, S. "The practice of radiology in the United States: facts and figures," Am. J. Roentgenol. **66,** 929 (1951).
19. Moeller, D., Terrill, J. and Ingraham, S. "Radiation exposure in the United States," Public Health Reports **68,** 57 (1953).
20. NCRP. National Council on Radiation Protection and Measurements. *Exposure of the U.S. Population from Diagnostic Medical Radiation*, NCRP Report No. 100 (National Council on Radiation Protection and Measurements, Bethesda, Maryland) (1989).
21. Johnson, D. and Goetz, W. "Patient exposure trends in medical and dental radiography," Health Phys. **50,** 107 (1986).
22. Kuhn, H. "Methods for reducing patient dose: rare-earth screens, filtration, spot film technique and digital radiography," Br. J. Radiol. **18** (Suppl.), 37 (1985).
23. NCRP. National Council on Radiation Protection and Measurements. *Mammography - A User's Guide*, NCRP Report No. 85 (National Council on Radiation Protection and Measurements, Bethesda, Maryland) (1986).
24. Cascade, P., Peterson, W., Wajszczuk and Mantel, J. "Radiation exposure to patients undergoing percutaneous transluminal coronary angioplasty," Am. J. Cardiol. **59,** 996 (1987).

Discussion

NAOMI HARLEY (NYU): Your slide on the Kienbock intrigued me and the penny dropped. Because this is probably the famous Adamson Kienbock procedure. And there were five shots per tinia capitis ringworm. And it was given left, right, front, top, back. So those probably are the five Kienbocks. And then at NYU, having done a lot of this reconstructive dosimetry, that's 350 rads.

FRED METTLER: That's great! Harold, do you have any comments on those units?

JOHN CAMERON (University of Wisconsin): About 10 to 15 years ago, a committee of the Bureau of Radiological Health, recommended unanimously that all x ray equipment in the United States have included with it a device to measure essentially the delivered roentgens per square centimeters where nowadays we would say the energy imparted in millijoules. This is still technically possible for a few thousand dollars per machine. Do you think in view of the large variations that this would be a good thing for NCRP to get behind, or some committee to consider whether we shouldn't be doing this for the largest man-made radiation sources?

FRED METTLER: Well, I think, certainly in my state, the state shows up every year and jumps on my head and wants to see all the calibration data for every machine. There's no question right now that the Joint Commission on Hospital Accreditation is requiring me to be able to produce an estimate of the dose for any given patient of whom I take an x ray. So my technologists, on every requisition, write the mA, the kV, what room it was done in, and so on. We have the calibration for that room. I am able to do that on any patient, on any x-ray machine and be patient specific. On the other hand, if I were to go across the street and open an office as a family practitioner, I could install anything, and the state doesn't come around and check. So I think the problem is that there are a lot of places out there that are really not checked by anybody. I think that's the biggest problem. Putting a meter on the machine I don't think is going to help because they're just not going to look at it. No matter what's going on, I think there needs to be a better regulatory process out there for those other groups.

SYDNEY PORTER (Porter Consultants): Fred, one historical correction, in 1899 Marie Curie in several publications suggested the use of radium for therapy treatment. And she started it early thereafter. In 1902, her good friend, Robert Abbey, a physician of Philadelphia and New

York, used the idea that he had gotten from Marie Curie and had cervical applicators, one of them, the 1903 edition of that, exists in Philadelphia today in the College of Physicians along with a lot of data showing it had been used for a number of years before 1903 in Philadelphia.

FRED METTLER: Yes, you know Sid, it's interesting, I went back, for example, to find out who took the first x ray in the United States. And there is a huge war. There are at least 15 people who claim that they were the first one who took an x ray. It's incredible, they're down to talking about days. But I grant you I did read some of those articles and it struck me that Alexander Graham Bell had sort of gotten involved in this.

SYDNEY PORTER: Marie Curie was inducted into the French Academy of Physicians in about 1902 or so. The first woman ever to be done so, and it was because of her suggestion of the use of radium for therapy.

OWEN MERRIL (Advisory Committee on Nuclear Waste, Staff Support Engineer): There was one bit of data that you didn't report on that I would like to pass on to you and anyone else, particularly Dr. Schull, who indicated in his summary that he would be speaking regarding lenticular opacities. I was born in 1926, while my father was interning or in residence at the LDS Hospital in Salt Lake City, and had developed what they thought was possibly a tumor on the mastoid area. I am told, but, I can't give you the values, that I was treated with very heavy doses of x ray. And as a result of that, I have a loss of bone structure in the left side of the face, the hair is finer, and there is less of it. I have developed a small cataract which finally matured when I got flown out of the High Sierras because I had suffered pulmonary edema and a lack of oxygen which I assume may have been a factor. And within two months after that I had total opacity of the left lens, which I had to have removed. So if you'd like to know more about it, or anyone else is interested in lenticular opacity, I'm an example.

JACK SHAPIRO (Harvard): Do you think it's productive at all for a hospital radiation protection person to work on trying to reduce prescription of x rays? That's been recommended in some of the reports. I noticed you had about a 1.2 factor if you worked on that. Do you do any of that at your own hospital?

FRED METTLER: That's a good question, Jack. You know I started that when I was at Massachusetts General when you and I were associating back in seventy. We had an interesting study that the BRH funded; which was looking at pregnancy and radiation. What we were able to do with our computer system was to track what doctors ordered what kind of x rays. We did that for about six months and got a profile on each doctor. Then I went around and yelled at them and screamed at them, about all the things they were going to get sued and so forth if they didn't stop ordering x rays on potentially pregnant women. Unbeknownst to them I then tracked them in the computer again. Their ordering rate on women in the child-bearing age group went way down for about three months. And then it went right back up again in about six months, and it's difficult to change these people's habits. Certainly every time we get a request for an x ray, and the hospital is mostly interested for financial constraint, I must say, rather than for x ray protection considerations. We get an x ray request says, rule out pneumonia. That may make sense if they've put a stethoscope on the patient and actually heard something or they were worried, but most of them have not even seen the patient by then. You say well what is the patient's trouble, but the physicians have not seen him yet. You don't know that when you meet the requisition. I think we have to make them think more about what they're doing, but it's a very tough and difficult process. Right now I'm just beginning to get DRG discharge diagnoses, and looking at the profile of a doctor on how many x rays he ordered on an appendicitis patient that went home. What I'm going to do is if I find an outlier, I'm going to go yell at him but I know it's only going to work for six months.

An Historical Perspective of Human Involvement in Radiation Accidents

C.C. Lushbaugh[1]
S.A. Fry
A. Sipe
R.C. Ricks
Oak Ridge Associated Universities
Oak Ridge, TN

Telling the truth about radiation accidents that have occurred since the dawn of the nuclear energy era can get you nowhere but somebody ought to do it—or try to. Such a person shouldn't expect, however, to get much help at all from the popular or even the scientific press media. The "press" of either focus must sell its product, and the truth about radiation accidents is incredibly dull, and obviously won't sell because actuality doesn't meet expectations. Even *"Science,"* the weekly magazine that is read by most of us who aspire to the truth about our physical, chemical, or biological environment or our bodies has an uncertain policy in this regard. Recently an Associated Press headline announced that a "survey suggests most Americans are illiterate about science." That statement is easy to believe because even if you read it backwards it is also true: "Science [is] illiterate about Americans." In a recent issue of *Science*, for example, a bold headline announces Professor Neel's study on human radiation genetics by stating "Lower Radiation Effect Found" (1). What does such a headline mean? I wonder how many

---

[1]The submitted manuscript has been authored by a contractor of the U.S. Government under Contract No. DE-AC05-76OR00033. Accordingly, the U.S. Government retains nonexclusive, royalty-free license to publish or reproduce the published form of this contribution, or allow others to do so for U.S. Government purposes.

readers were disappointed to read on and learn that the headline writer had assumed that the reader would anticipate that Professor Neel found five times less radiation-induced genetic damage in man than the mouse model on which our guidelines are based would suggest. Why not write "Human found to be five times more radioresistant than Mouse!? (No one would believe it!) After reading in a recent lead article in *Science* by Anspaugh, Catlin, and Goldman (2) that 50,000 persons *had received exposures of 200 rads of radiation* in the Chernobyl accident, one of us had the temerity to point out this apparent overstatement to the Science Letters Editor so that the word "committed," could be inserted. A prompt reply, however, informed us that if our letter was ever needed "they" would let us know. It wasn't ever "needed," however, because one of the authors wrote a blind "answer" to our query stating that "dose" (not "committed dose") was actually intended(!). In so stating, the author indicated that many of these 50,000 instances had *accumulated* the 2 Gy "dose" slowly over as much time as a year during the post Chernobyl cleanup! So, these 50,000 "doses" were not "acute" and must have been "accumulated" rather than "committed" doses. Our purpose in dwelling on this "media event" is to make the point that health physics "experts" may be promoting the scientific illiteracy of the American public and helping the media, technical as well as "popular," to confuse the "public." In Table 1 we've listed some familiar "doses" that could be confusing to the general "public" who, as cartooned in Figure 1, probably believe that as stated in the famous poem about a "rose (being) a rose," *a dose is a dose!* If so, it follows naturally that *any* radiation dose is an unsafe dose and as shown in Figure 2, the life expectancy of nuclear power is considered to be worse than uncertain because the "patient" is terminal!

As shown in Figure 3, *Time* magazine may have seen the light recently since it now asks the question, "Is Anything Safe?" Surely the editors of *Time*, if not the average man or woman, know that being "mortal" means one has a great certainty of dying. So the answer to *Time's* question is "No!" although some might answer instead that "some things are safer than others." The 45 year long survey of worldwide radiation accident history that we are about to review suggests, judged from the incidence of radiation accident fatalities, that working with, in, or around radioactivity of all conceivable kinds has been incredibly safe in comparison with other activities as shown in Table 2.

Table 2 provides the most recent list of causes of accidental deaths in 1985 in the U.S. that has been provided by the National Safety Council (3). The 79,057 deaths on this list did not involve radiation exposure in any way. A more complete list of all deaths due to injury (pp 12–13, ref

TABLE 1—*"Dose" As a Scientific (?) Literacy Problem for Amateur Health Physicists and the Average American*

Dose

Air Dose

Exposure Dose

Depth Dose

Absorbed Dose

Estimated Dose

Committed Dose

Accumulated Dose

Prompt Dose

Equivalent Dose

Cytogenetic Dose

Relevant Dose

Organ Dose

Critical Organ Dose

Residual Dose

Lethal Dose

LD 90/60

LD 50/60

LD 10/60

(Etc. But Don't Mention "Units")

3) shows that no deaths due to radiation occurred in the U.S. during 1983, 1984, and 1985. But what about the incidence of radiation deaths in the U.S. and the rest of the world since the dawn of nuclear power in 1944 about 45 years ago? The REAC/TS Radiation Registry reported here follows the simplistic rule that *all* deaths must be counted if they occurred during or shortly after a radiation accident even if the death was sec-

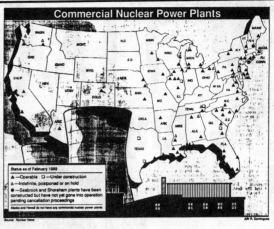

Fig. 2.

ondary to non-radiation trauma (like thermal burn or blast injury) or even if the death would not have occurred had proper (?) treatment been administered. Such uncensored data reveal that the total number of fatalities is 101.

Since 1974 in the DOE/Oak Ridge Operations' Radiation Emergency Assistance Center/Training Site program, a Worldwide Radiation accident registry has been kept in great detail.[2] Previously, several published reports (4–9) have documented the record and accumulation of radiation accidents along with radiation deaths and deaths from associated causes. Figure 4 shows in detail the number of separate registries, the number of registered radiation accidents, and the number of persons involved in these events. Of the 986 total "accidents," 587, which involved 1782 persons, were "non-accidents" in which no radiation exposures occurred. There were only 305 worldwide bonafide radiation accidents—188 in the U.S. and 117 in the rest of the world. This apparent disparity may be the result of a more complete surveillance of radiation therapeutic misadventures and better reporting of such "accidents" in the U.S. than in other countries (except perhaps for United Kingdom, France, West

---

[2]Present (1989) Registrar, Dr. S.A. Fry, Associate Registrar, Ms. A. Sipe, LPN, Program Director, Dr. R.C. Ricks.

Fig. 3.

TABLE 2—*Accidental Deaths Causes in the U.S. 1985*

45,901	Motor vehicle
12,001	Fall
4,938	Fire
4,407	Drowning
3,612	Drug or medication
1,663	Food inhalation or ingestion
1,649	Firearms
1,428	Air travel
1,288	Machinery
903	Struck by falling object
802	Electric current
305	Alcohol poisoning
85	Lightning
49	Venomous bite or sting
15	Dog bite
11	Fireworks

Source: National Safety Council

Germany, Italy, and (recently) P.R. China). Even so, the number of known radiation accidents worldwide over the last 45 years remains comparatively small, and the total numbers of individuals involved in them so remarkably few that they are commonly explained as pro-nuclear propaganda. The 1986 radiation accident in Chernobyl USSR (10–11) and the 1987 Goiania, Brazil, (12–13) accident increased these figures as shown in Figure 5 and added 36 fatalities to the 65 recorded over the last 45 years. In this figure, the statistics "Persons Involved" and "Significant Exposures" are displayed as numbers with potential utility for assessing the use of radioactive materials more critically or less encouragingly. The figure (Fig. 5) shows that up until Chernobyl only 5,865 persons were directly involved in 303 accidents wherein only 1,335 persons received radiation exposures of more than 0.25 Gy (total body) or more than 6 Gy to a skin area large enough to produce local symptoms. As far as we can determine from open literature reports by the USSR physicians (14), there were about 116,500 persons involved (these persons were evacuated from the 30 kilometer zone and include 499 to 500 persons with acute dose of $\geq$ 0.25 Gy total-body irradiation who were of immediate medical concern and transported to the Kiev and Moscow hospitals. The 237 persons reported to have received acute exposure greater than 0.8 Gy were those transported to Moscow hospital #6 for intensive care (10, 14). The 50,000 persons recently reported by Anspaugh *et al* (2) to have been exposed to 2 Gy in the year following the accident (apparently during the environmental clean-up) casts doubt on the accu-

Status of REAC/TS Registries[a]

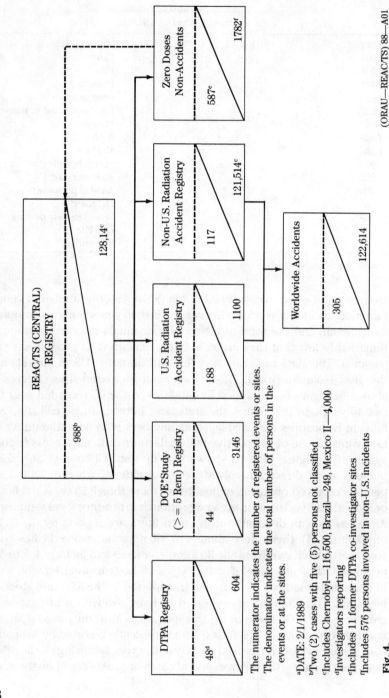

The numerator indicates the number of registered events or sites.
The denominator indicates the total number of persons in the events or at the sites.

[a]DATE: 2/1/1989
[b]Two (2) cases with five (5) persons not classified
[c]Includes Chernobyl—116,500, Brazil—249, Mexico II—4,000
[d]Investigators reporting
[e]Includes 11 former DTPA co-investigator sites
[f]Includes 576 persons involved in non-U.S. incidents

Fig. 4.

(ORAU–REAC/TS) 88—A01

Major Radiation Accidents Worldwide Human Experience
1944—February, 1989[a]

Number Accidents	Persons Involved	Significant Exposures	Total Fatalities*
303	5,865	1,335	65
(1	116,500	500[b]	32)
		237[c]	
[1	249	36	4]
305	122,614	1,871	101

[a]Source: DOE—REAC/TS Radiation Accident Registries
[b]USDOE/NRC Accident Dose Criteria
[c]> = 80 Rads
( )—Chernobyl Data
[ ]—Goiania Brazil Data
*Includes deaths due to acute radiation effects and deaths from other trauma
(ORAU—REAC/TS) 88—007

Fig. 5.

racy of our number of "Significant Exposures" and suggests that our concepts here need to be reviewed. The Brazil accident statistics are less complicated at first glance in that only 249 persons were "involved" and only 36 were found to have had skin contamination and total-body and skin radiation exposures that required medical intervention (hence "significant exposures"). In both the Chernobyl and Brazilian accident, there were large numbers of contiguous unirradiated persons who suffered from radiation phobia: in Brazil 112,000 presented themselves voluntarily because of contamination that was never found. "Radiation phobia" is undoubtedly a real problem that suggests another criterion by which "involvement" in a radiation accident might be measured. Up until now, however, the REAC/TS registry of radiation accidents and their medical effects has been limited to recording observable physical damage and measured exposures usually delivered promptly but, in a few instances, over as much as 100 days. With these caveats, the Radiation Accident Registry now contains 305 accidents in which 122,614 persons were involved, of whom 1,871 received "significant" exposures, during or after which 101 total fatalities occurred (not all of which were due to radiation tissue damage—see below).

In Figure 6, these 305 radiation accidents have been listed according to the "device" or the kind of radiation usages involved. There appears from this list to be three main causes of accidents: criticality, radiation-emitting devices, and radioisotopes. The accidents involving diagnostic and therapeutic uses of radioactive materials and radiations are included

Major Radiation Accidents: Worldwide
1944—February, 1989

Classification of Radiation Accidents by 'Device'

Criticalities[a]	
Critical Assemblies	5
Reactors	8
Chemical Operations	5
Radiation Devices	
Sealed Sources	140
X-Ray Devices	62
Accelerators	14
Radar Generators	1
Radioisotopes	
Transuranics	27
Tritium	1
Fission Products	10
Radium Spills	2
Diagnosis and Therapy	24
Other	6
Total	**305**

[a]Only 2 since 1965

(ORAU—REAC/TS) 88—003

Fig. 6.

under "Radioisotopes." The 14 accelerator accidents are considered "industrial mishaps" and for this reason included under "Radiation Devices" even though their use in radiation therapy might suggest otherwise. Obviously, the most frequent cause of radiation accidents comes from the misuse of "sealed sources," shown here as 140 of the 305 total accidents. This remarkable frequency is illustrated well in Figure 7, a bar graph depicting the frequency of these three major classes of accidents per 5 year periods from 1950 to date. The increasing frequency of accidents from RD (Radiation Devices) until 1979–1980 suggests that these accidents were related to the necessity for oil exploration and their subsequent decline followed the oil "glut" that then developed. An alternative explanation is that the redesign of the equipment by NRC helped prevent these radiation devices becoming loose in the environment. This figure also shows that criticality accidents stopped occurring after 1970 when remote control and automation were developed and occurred again in 1984 (in Argentina in a research reactor) and again in 1986 (in Chernobyl); two accidents in the real sense of the term. This graph also shows that over this 45-year period radiation accidents from

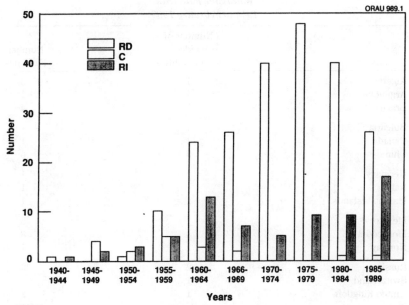

Fig. 7.

the well-intended use of radioisotopes (RI) has been constant over time at an average rate of about 2/year—a remarkably low rate when the worldwide everyday use of radionuclide applications is considered (or when industrial or automobile accidents are compared).

In Figure 8, we attempt to "point the finger" at the countries where fatalities have occurred as the result of radiation accidents. This tabulation shows that of the 305 radiation accidents in 45 years, only 33 have been attended by fatalities; 14 of these have been in the U.S. and in the USSR. A rundown of the countries of occurrence seems meaningless and a matter of chance rather than the result of usage or lack of safety procedures; except perhaps for U.S. We suggest that in this instance we (REAC/TS) as a highly motivated U.S. institution are more efficient in detecting and recording U.S. accidents than those occurring worldwide. This bias is reflected in many parameters in Table 3 which details many of the parameters of the data underlying the 101 fatal radiation accidents which we have recorded. Here we have indicated the year, the place and the number of persons involved and then shown the number (who having met our dose criteria) were of medical concern. Then we have divided, as best we could, the actual deaths into three subsets: Acute Radiation Syndrome deaths; delayed (>60 day) or associated deaths (deaths where

Worldwide Fatalities
1944 to February 1989

Country	Number of Accidents with Fatalities	Number Fatalities
Algeria	1	2
Argentina	1	1
Brazil	1	4
Bulgaria	1	1
Canada	1	1
China	3	6
Germany	1	1
Italy	1	1
Marshall Islands	1	1
Mexico	1	5
Morocco	1	8
Norway	1	1
Russia	2	33
Switzerland	1	1
United Kingdom	1	2
United States	14	32
Yugoslavia	1	1
Total	33	101

Fig. 8.

radiation exposures were involved and would have otherwise not have occurred) and non-radiation deaths due to trauma from nonradiation energies or nonradioactive chemicals released in a radiation accident. The data are admittedly difficult to analyze but when assessed from the point of view of the question of "how many of the 101 total fatalities in radiation accidents were due to radiation?" the best answer seems to be only 50 in 45 years. Deaths from blast, thermal burn, or hydrogen flouride numbered 11. Deaths from radiation therapy, nuclear medical procedures, and medical misadventures numbered 28. Seven deaths were delayed, three were suicidal rather than radiation accidents, and two were due to uncontrolled chronic occupational exposures. When these 40 radiation-associated deaths (?) and the 11 non-radiation deaths are subtracted from the 101 total fatalities in radiation accidents, the best answer seems to be that there have been only 50 acute radiation accidental deaths since the development of nuclear energy began.

TABLE 3—*Fatal Radiation Accidents*

Year	Place	Number Involved	Met Dose Criteria	Acute*	Number of Deaths Delayed or assoc.	Non Rad**	Total
44	Phil, PA	20	4	—	—	2 F	2
45	LA NM	2	1	1 BM	—	—	1
46	LA NM	8	8	1 BM	—	—	1
54	Marshall Islands	290	133	—	1 H	—	1
58	Yugoslavia	6	6	—	1 BM	—	1
58	LA NM	3	3	1 VS	—	—	1
60	USSR	1	1	—	1 1	—	1
61	Idaho Falls	20	7	—	—	3 BL	3
61	Switzerland	3	3	—	1 BM	—	1
62	Mexico	6	6 (Fe)	1 BM	4 BM (Fe)	—	5
63	P.R. of China	6	6	2 BM	—	—	2
64	Rhode Island	7	4	1 VS	—	—	1
64	Germany	4	4	—	1 BM	—	1
68	Wisconsin	1	1	—	1 M	—	1
72	P.R. of China	15	15	2 BM	—	—	2
72	Bulgaria	1	1	—	1 1	—	1
75	Columbus, OH	403	403	—	10 M	—	10
75	Italy	2	1	1 BM	—	—	1
78	Algeria	8	8 (Fe)	2 BM (Fe)	—	—	2

183

TABLE 3—continued

Year	Place	Number Involved	Met Dose Criteria	Acute*	Number of Deaths Delayed or assoc.	Non Rad**	Total
80	Houston, TX	8	8	—	7 M	—	7
81	Oklahoma	1	1	—	1	—	1
82	Norway	1	1	1 BM	—	—	1
83	Argentina	18	1	1 VS	—	—	1
84	Morocco	26	9	8 BM	—	—	8
85	P.R. of China	3	3	2 BM	—	—	2
85	Canada	1	1	—	1 M	—	1
86	Tyler, TX	1	1	—	1 M	—	1
86	Tyler, TX	1	1	—	1 M	—	1
86	USSR	116,500	500	22 BM	3 BM/2M	1 Bl/4 BU	32
86	Gore, OK	110	1	—	—	1 F	1
87	Yakima, WA	1	1	—	1 M	—	1
87	Brazil	249**	36?	4 BM	—	—	4
88	UK	207	207	—	2 M	—	2
		117,933 ?	138??	50	40	11	101

*BM = ARS (Bone Marrow)   H = Hepatitis   Fe = Includes fetal death   ***112,800 monitored   30 Medical
VS = Vascular syndrome   M = Medical misadventure   Bl = Blast   249 "contaminated"   surveillance
I = Infection (2°)   **F = HF₆ Gas   Bu = Burn   129 int. and ext. contamination   20 hospitalized

Fifty acute radiation deaths are, of course, too many to accept upon the basis of humanitarian principals. When divided by 45–50 years, however, this number of deaths becomes a death rate of about 1/year/ >200,000,000 persons, a number too small ever to be achieved by any other professional occupation in the world, either east or west.

References

1. Marx, J.L. "Lower radiation effect found," Science (News Article) **241**, 1286 (1988).
2. Anspaugh, L.R., Catlin, R.J., Goldman, M. "The global inpact of the Chernobyl reactor accident," Science **242**, 1513–1519 (see p. 1517, paragraph 7) (1988).
3. Accident Facts, National Safety Council, 1988 edition (ISBN: 0-87912-139-4).
4. Lushbaugh, C.C., Fry, S.A., Hubner, K.F. and Ricks, R.C. "Total-body irradiation: an historical review and follow-up," page 3–15 in *The Medical Basis for Radiation Accident Preparedness* Hubner, K.F. and Fry, S.A. (Eds.) (Elsveir North Holland, New York) (1980).
5. Fry, S.A., Lushbaugh, C.C. and Hubner, K.F. "An overview of radiation accidents and injuries," Int. J. Nucl. Med. Biol. **9**, 152–153 (abstract) (1982).
6. Lushbaugh, C.C., Fry, S.A., Ricks, R.C., Hubner, K.F. and Burr, W.W. "Historical update of past and recent skin damage radiation accidents," Brit. J. Radiol. Sup 19, page 7–12, London (1986).
7. Lushbaugh, C.C., Fry, S.A. and Ricks, R.C. "Medical radiobiological basis of radiation accident management," page 1159–1163 in *Proceedings of the British Institute of Radiology's 1987 Annual Congress*, April 2, 1987, Southampton, England, Brit. J. of Radiol. **60**, 720 (1987).
8. Lushbaugh, C.C., Ricks, R.C. and Fry, S.A. "Radiological accidents—an historical review of sealed sources accidents," in *Proceedings of IAEA International Conference on Radiation Protection in Nuclear Energy*, April 18–22, Sydney, Australia, October 1988 (International Atomic Energy Agency, Vienna)
9. Lushbaugh, C.C. "Radiation accidents world-wide: An additional decade of experience," in *Proceedings of the Second REAC/TS International Conference "The Medical Basis for Radiation Accident Preparedness: II Clinical Experience and Follow-up since 1979*, October 20–22, 1988 (in press).
10. Guskova, A.K., Baranov, A.E., Barbanova, A.V., Gruzdev, G.P. and Piatkin, E.K. "Acute radiation effects of the victims of the Chernobyl nuclear power station accident," Med. Radiology (Moscow) **32**, 3–18 (1987).
11. Baranov, A. "The acute bone marrow syndrome in the Chernobyl accident, 1986," in *Proceedings of the Second REAC/TS International Conference "The Medical Basis for Radiation Accident Preparedness: II Clinical Experience and Follow-up since 1979"*, October 20–22, 1988 (in press).

12. Oliveira, A.R. "Skin lesions associated with radiation exposure and contamination: Goiania, Brazil," in *Proceedings of the Second REAC/TS International Conference "The Medical Basis for Radiation Accident Preparedness: II Clinical Experience and Follow-up since 1979"*. October 20–22, 1988 (in press).
13. Valverde, N.J. "The acute radiation syndrome in the 137-Cs Brazilian Accident, 1987," in *Proceedings of the Second REAC/TS International Conference "The Medical Basis for Radiation Accident Preparedness: II Clinical Experience and Follow-up since 1979"*. October 20–22, 1988 (in press).
14. Pyatak, O.A. All-Union Scientific Centre of Radiation Medicine, Kiev, USSR (Personal Communication).
15. McGuire, S.A. and Peabody, C.A. *Working Safety in Radiography*, NUREG/BR-0024 (U.S. Nuclear Regulatory Commission, Washington) (1982).

Discussion

ROBERT VAN WYCK (Long Island): Just a comment. I'm reminded this afternoon of the pleasure of working with Charles Lushbaugh on the now defunct subcommittee 13. And I'm reminded of the pleasure of those meetings. And this is perhaps why we kept meeting and meeting and meeting and meeting. And from time to time we would get some work done. But they were totally enjoyable. But this was also highly informative. I'd also like to conclude on a somewhat up-beat note for the day in that I'm from Long Island, the home of the would be Shoreham Power Plant. A few weeks ago, there was a newspaper poll conducted where you phone in yes or no. Sixty-five percent of the respondents voted for the plant to operate at full power.

ALAN TANNER (U.S. Geological Survey): The U.S. Bureau of Mines quoted to me last year that the 1987 coal mine deaths in the U.S. were sixty-four. And that's the lowest number in history.

MARVIN GOLDMAN (University of California): I couldn't resist trying to attempt to come up with something in response to your historic enrichment on the Chernobyl data that we published in Science. There is some confusion, and many of you brought up some of the points about that, and I'd like to clarify it. With regard to the 116,000 people for whom we talked about committed doses, those indeed are the doses they received and those that are projected forward that the evacuation population might receive in the future from contaminated food and the environment. In addition to that, what few people are aware of is the fact that they brought in some 200,000 workers, many of whom are in the military, who received very large doses, some of whom were perhaps a cohort of about 4000 people receiving 200 rads and another 10,000 receiving 100 rads. Those were absorbed in that one year. We attempted to explain all of that in the Science paper but the editors insisted that we make this 300 page report smaller so that they could publish it. And so I did specifically use the word committed dose where it was appropriate and absorbed dose where appropriate. And they would not allow me to elaborate further on it. I suspect the list of ways of handling dose is even longer but we can talk about that tomorrow.

AL TSCHAECHE (WINCO): Dr. Lushbaugh, I think I heard you say that in the Goiania accident there were some people who were contaminated who somebody decided didn't have to be decontaminated and that there was a level at which they had contamination, perhaps skin contamination, and they didn't bother to decontaminate them. Do you know what that level was?

C.C. LUSHBAUGH: It was a very practical point in that the persons who didn't need decontamination had contaminated clothing which they just had to remove. The other persons who were decontaminated had contaminated skin that had to be decontaminated. But of the total number of persons that had their skin decontaminated, there were only a few of them, I think my number says it was 129, that had to be very carefully decontaminated because they did have a contamination level that would have given them a significant committed dose to the skin. Those persons were decontaminated.

ED WRENN (University of Utah): I really enjoyed your table that showed the different sources of death. I may have seen something in there that others didn't. I saw a negative death toll from Americium-241 of 3800 people per year. Now how can Americium produce a negative death toll. If you go back fifteen years, go pick a National Safety Council, you'll see 8800 deaths from fires a year. And that one had just under 5000. So that's 3800 people; and that's the smoke detector. And ninety percent of those are fueled with Americium-241.

Lauriston S. Taylor Lectures in Radiation
Protection and Measurements

Lecture No. 13

Radiobiology and Radiation Protection: The Past Century and Prospects for the Future

by Arthur C. Upton

Presented April 5, 1989
Issued July 1, 1989

National Council on Radiation Protection and Measurements
7910 WOODMONT AVENUE/BETHESDA, MD. 20814

Lauriston Sale Taylor

THE LAURISTON S. TAYLOR LECTURES
IN RADIATION PROTECTION AND MEASUREMENT

It is my pleasure to welcome you all to the Thirteenth Lauriston S. Taylor Lecture in Radiation Protection and Measurement.

These lectures are intended to honor Lauriston S. Taylor who was the first President of the National Council on Radiation Protection and Measurements (NCRP) from its inception in 1964 until he retired in 1977. Prior to 1964, he was chairman of all the NCRP's predecessor national committees beginning in 1929. A detailed account of his career and achievements is contained in an appreciation which appears in Lecture No. 1 of this lecture series.

The lectures are given once annually by distinguished lecturers selected by the program Committee of the NCRP. They treat a topic related to radiation protection or measurement. The lectures are sponsored by the NCRP, are held on the occasion of its Annual Meeting and are published subsequently under the auspices of the Council.

Initiation of the lecture series was made possible by a generous grant from the James Picker Foundation. For this the Council expresses its deep appreciation.

The NCRP is sixty years old this year. We are delighted to have both Lauriston Taylor and his wife, Robena, here today. It must be rare indeed for a man to attend the sixtieth birthday of the organization he founded. Laurie (and Bena) I would like you to stand and I would ask the audience to show their appreciation by a round of applause for Laurie and Robena Taylor.

The Thirteenth Lauriston Taylor lecturer, Dr. Arthur Upton of New York University, will be introduced by his distinguished friend and colleague, Dr. Naomi Harley, also of New York University, herself a well known worker in the fields of dosimetry and epidemiology and a prominent member of the NCRP. Dr. Harley. . . .

Warren K. Sinclair
President, NCRP

Arthur C. Upton

ARTHUR C. UPTON

*Thirteenth Lauriston S. Taylor Lecturer in
Radiation Protection and Measurements*

Today we honor one of the most distinguished persons in the field of radiation protection, Lauriston S. Taylor. It is fitting that the 13th Taylor Lecturer be a man whose life is dedicated to the same principles as Laurie's. It is naturally a great honor to introduce the Taylor Lecturer and I am personally pleased because Arthur Upton is the man for whom I work and who shapes my scientific environment.

Arthur C. Upton was born in and spent much of his early life in Ann Arbor, Michigan. He was literally wheeled into academic life as his grandfather would put him on the handlebars of his bike and take him to the campus of University of Michigan where grandpa was chairman of the Department of Romance Languages.

His strong interest has always been in medical research. At the age of 10, his mother fell ill with pneumonia, then a serious and life-threatening disease. The physician treated her with sulphonamides and Arthur was impressed by her recovery. Arthur wanted to be like this man who could actually do something to help the course of disease.

His undergraduate work and medical studies were at the University of Michigan during World War II in the armed services training program. He graduated from medical school in 1946 as a US Army Captain. During his studies, his parents asked him what he planned to do in medicine. He said he did not know but one thing he would not do would be to go into pathology. But the call of research was too strong and pathology was the field that he felt led to better understanding of disease. He did his internship and residency and became an instructor in pathology at the University of Michigan.

His first publications (of over 200) emerged in 1950.

In 1954 Arthur was seeking a research position when Jacob Furth from Oak Ridge came looking for a young research pathologist. This happy combination started Arthur in the path of righteousness, namely radiation protection. He and his wife Betsy moved to Oak Ridge National Laboratory where he became chief of the pathology—physiology section.

During their years at Oak Ridge, Betsy and Arthur were also raising a family of three children Rebecca, Melissa and Bradley. Betsy is a doctor in her own right, doing graduate work at the Univ. of Tenn. while she was at Oak Ridge and later at NYU, attaining a PhD is Spanish. Melissa is the one who follows in her father's footsteps and is a pathologist and researcher at the Boston VA hospital.

Richard Strauss in his opera Ariade auf Naxos said "there is a realm where all is pure—this realm is death." Arthur has in a sense devoted his life to study death with the view that such purity should be postponed and human life prolonged.

One of his first works in radiation biology was a study to show the protective effects of methylcholine against x-ray radiation.

The opportunity to return to the academic world lured him to Stony Brook, then a new campus and medical school on Long Island in 1969. The prospect of collaborating with researchers at Cold Spring Harbor and Brookhaven was overpowering. He was soon asked to become the Dean of the School of Basic Sciences. This took him out of his favorite, research, but he felt it was important to see the Department develop.

In 1977 he moved to Washington to become the Director of the National Cancer Institute where he was a captive of administrative responsibility. Wanting to return to academia he became the Director of the Institute of environmental Medicine and Chairman of this Department at New York University School of Medicine in 1980, when Norton Nelson stepped down from this position.

Since the age of 27 he has held responsible positions in academia and the National Laboratories.

Reviewing Arthur Upton's acomplishments is inspiring. He has been awarded over 20 honors in cancer research, oncology and environmental science. He has held office in many national societies, becoming president of several, for example, the American Association for Cancer Research, the American Society for Experimental Pathology and the Radiation Research Society—to name a few.

Arthur has chaired or served on an unbelievable number of committees, giving guidance to such varied groups as the National Academy of Sciences (he is now chairing BEIR V), the Japan Radiation Effects Research Foundation, WHO, the Three Mile Island Unit 2 Safety Advisory Board, the Lawrence Livermore Review panel on malignant melanoma and a list of 100 or so more. Truly Arthur Upton is currently instrumental in shaping the world of radiation and health.

In my working relations with Arthur I know him as a man of patience and tolerance worthy of great respect. His smile is a hallmark. He accepts his staff and others as they are with their shortcomings and works in a positive way to emphasize the pluses and ignore the minuses.

Peter Cooper founded The Cooper Union, my own undergraduate alma mater. Peter Cooper's motto was "the object of life is to do good". Nowadays many turn the noun into the adverb, making it "the object of life is to do well". Arthur exemplifies the motto of "to do good" in his career and in his personal life. Science and medicine have been the happy recipients of this attitude.

His great virtue aside from his own research, is the ability to help and encourage people in the field to which he has devoted his life and to move the work along. In doing this he has won the esteem of people around the world. The book of Proverbs says, "to be esteemed is more precious than silver or gold" and esteem Arthur has in abundance from colleagues, fellow physicians and scientists and I am sure, by Lauriston Taylor, whom we honor today.

It is my great pleasure to introduce this unique scientist, Arthur C. Upton, to deliver the 13th Lauriston S. Taylor

Naomi H. Harley

Lecture entitled "Radiobiology and Radiation Protection: The Past Century and Future Prospects".

>Naomi H. Harley
>New York University Medical Center
>New York, New York

The Lecturer: Arthur Upton

Radiobiology and Radiation Protection: The Past Century and Prospects for the Future.[1]

by ARTHUR C. UPTON

DEDICATION

I am pleased and honored to have been invited to present the 13th Annual Lauriston S. Taylor Lecture on Radiation Protection and Measurements. Having devoted my professional life to the study of radiobiology and its applications to radiation protection, I am thankful to have known Laurie Taylor as a friend and to have worked with him and the organization that he created—the NCRP—for more than 30 years.

Anyone acquainted with the history of radiation protection is well aware of Dr. Taylor's unique contributions to the

---

[1]Preparation of this report was supported in part by Grants ES 00260 and CA 13343 from the U.S. Public Health Service and Grant 8-0248-302 from the American Cancer Society.

field, beginning with his Chairmanship of the country's first Advisory Committee on Radiation Protection, in 1929, thereafter continuing without interruption for nearly half a century—during which the Committee evolved under his chairmanship into the NCRP—and continuing since then to the present time. No one has played a more central role in the development of the philosophy and practice of radiation protection than Lauriston Taylor.

From the outset, his vision and leadership have been instrumental in enlisting experts from the different disciplines whose special talents and combined efforts have been required for addressing the diverse problems encountered in this rapidly developing field. The results of their combined efforts have guided radiation protection activities the world over, to the benefit of people everywhere.

With appreciation for the personal pleasure, satisfaction, inspiration, and intellectual growth that I have experienced in working with Laurie and the NCRP, as well as in recognition of Laurie's outstanding contributions to the entire field, I am pleased to have this opportunity to dedicate my remarks to him.

THE BEGINNING OF RADIOBIOLOGY

In reviewing the progress that has been achieved in radiobiological research, one cannot help but be impressed by the rapidity with which the field gathered momentum in the first years of its existence. Within months after the discovery of the X-ray by Roentgen, in 1896, radiation became introduced widely into the diagnosis and treatment of disease, with the result that injuries began to be encountered in radiation workers almost immediately (1,2). Among the first to experience such effects were: Pierre Curie—who intentionally exposed the skin of his own arm to observe its reaction—Marie Curie—who developed radiation burns on her fingers as a result of handling a small tube of radium enclosed in a thin metal box—and Antoine Henri Becquerel—who burned the skin on his chest by carrying a tube of radium-bearing barium chloride in his vest pocket (3).

Table 1: Noteworthy Reports of Radiation Injury in the Early Period Following Roentgen's Discovery of the X-Ray, in 1896[a]

Date	Type of Injury	Authors
1896	Dermatitis of hands	Grubbé
1896	Smarting of Eyes	Edison
1896	Epilation	Daniel
1897	Constitutional symptoms	Walsh
1899	Degeneration of blood vessels	Gassman
1902	Cancer in X-ray ulcer	Frieben
1903	Bone growth inhibited	Perthes
1903	Sterilization produced	Albers-Schönberg
1904	Blood changes produced	Milchner & Mosse
1906	Bone marrow changes demonstrated	Warthin
1911	Leukemia in five radiation workers	Jagié
1912	Anemia in two X-ray workers	Bélére

[a]From 6.

The injuries that were noted initially consisted predominantly of skin reactions on the hands of those working with early radiation equipment and sources (4). Within barely a year after Roentgen's discovery, 96 cases of such injury had been reported in one publication alone (5). In less than a decade, many types of radiation injury were observed, including the first cancer attributed to irradiation (Table 1).

Since these early findings, nearly a century ago, study of the biological effects of radiation has received continuing impetus from the growing uses of radiation in medicine, science, and industry, as well as from the military and peaceful applications of atomic energy. As a result, the effects of ionizing radiation have been investigated more thoroughly than those of any other environmental agent.

The evolution of radiobiological knowledge and its application to radiation protection have ramifications of far-reaching scientific and practical importance. It is fitting, therefore, on the occasion of NCRP's 60th anniversary to review these developments and their implications for the future.

ELUCIDATION OF RADIOBIOLGICAL EFFECTS AND THEIR DOSE-RESPONSE RELATIONSHIPS

Tissue Reactions

By the dawn of this century, the reactions of different tissues to irradiation were under intensive study. Reactions of the skin were the first to be documented in detail, but reactions of other tissues also were investigated. As early as 1903, for example, Albers-Schönberg reported sterilizing effects of x-rays on male guinea pigs and rabbits (7), and the testes of the animals were found to be atrophic on histological examination (8).

There soon followed a series of further studies of the effects of x-rays on the testes, including those of Bergonié and Tribondeau (Table 2). The low radiosensitivity of mature spermatozoa, as compared with their precursor cells, led Bergonié and Tribondeau to the important generalization since known as the "law of Bergonié and Tribondeau". At about the same time, Regaud's studies disclosed that the sterilizing effects of gamma rays on the testes were paradoxically enhanced by fractionation—owing, we now know, to the relatively small percentage of stem cells that are in the most radiosensitive stage of spermatogenesis at any one time (22).

The early experimental studies of testicular reactions were paralleled by complementary clinical reports. In 1905, for example, Brown and Osgood described 18 cases of oligospermia and azoospermia in radiation workers, noting that the azoospermia was complete in those who had worked extensively with x-rays for more than three years (15).

The responses of the different tissues of the body to irradiation, and the ways in which they vary in relation to the dose and duration of exposure, have since been well characterized in radiotherapy patients as well as in laboratory animals. Curves characterizing such dose-response relationships became available for many organs of the body decades ago (e.g., Figure 1). The fact that susceptibility to radiation injury varies among different tissues, and depends also on

Table 2: Some Historical Highlights in Research on the Radiobiology of the Testis

Year	Observation	Investigator
1903	Azoospermia in x-irradiated guinea pigs and rabbits	Albers-Schönberg (7)
1903	Atrophy of seminiferous tubules in x-irradiated guinea pigs and rabbits	Frieben (8)
1904	Atrophy of testes of radium—exposed animals	Seldin (14)
1904–06	Variation of radiosensitivity in relation to degree of maturation of spermatogenic cells	Bergonié & Tribondeau (9–12)
1905	Azoospermia and oligospermia in heavily irradiated radiation workers	Brown and Osgood (15)
1906	Radioresistance of interstitial cells	Villemin (16)
1906	Radiosensitivity of spermatogonia; radioresistance of secondary sexual characteristics	Regaud and Blanc (13)
1922	Time-intensity relationship for effects on the testes of the ram	Regaud (17)
1927	Dose-response relationship for castration in the rat	Ferroux and Regaud (18)
1964	Comparative responses in different species	Oakberg and Clark (19)
1967	Quantitative dose-response data for human testes	Heller (20)
1971	Spermatogonial stem-cell renewal	Oakberg (21)

such factors as growth rate gave rise to two concepts which have been influential in radiation protection.

The first is the concept of a *tolerance dose;* that is, the concept that each tissue can withstand irradiation without significant reaction so long as the cumulative dose to the tissue does not exceed the relevant threshold (25). This concept remains the basis for protection against *nonsto-*

Arthur Upton

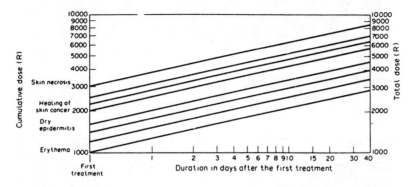

Figure 1 Estimates of exposure for equivalent damage to human skin in relation to duration of daily fractionated irradiation (From 23; reproduced from 24).

chastic effects of radiation (26). The dose limits are subject to continuing review, however, (27, 28), and major uncertainties at this time concern the thresholds for chronic high-LET irradiation, human data for which are sparse (26).

The second of the aforementioned concepts is the *critical organ* concept; namely, the concept that adequate protection of the entire body can be achieved by appropriately limiting the dose to its most radiosensitive and/or heavily irradiated tissue (29). Although this approach may be effective under certain conditions, it does not provide for the summation of risks to all organs from multiple exposures. Consequently, it has been superseded by a system that attempts to do so insofar as possible within the limits of present knowledge (30).

Effects on Cells

Modern mammalian radiobiology can be considered to date from the experiments of Puck and his associates, who demonstrated the feasibility of quantitative dose-survival studies with individual clonogenic human cells. The unexpectedly low D_o of the dose-survival curve they obtained (Figure 2) and their interpretation that the curve's initial shoulder represented a multihit inactivation process led ultimately to Elkind's studies demonstrating the intracellular repair of sublethal radiation damage (32). These findings and

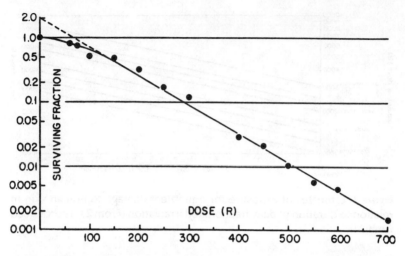

Figure 2 Survival of reproductive ability in isolated HeLa cells, in relation to dose of x-rays (Reproduced from 31).

the subsequent research that they stimulated have revolutionized our concepts of the response of mammalian cells to radiation injury. Although the high radiosensitivity of dividing cells had been known since the time of Bergonié and Tribondeau, the new approaches have explained the reactions of the various tissues in terms of differing changes in cell renewal kinetics, thus explicating dose-response relationships hitherto obscure (22, 33, 34).

The new techniques have also clarified the influence of LET on the dose-survival relationship (e.g., Figure 3), with far-reaching implications for risk assessment and radiation protection (22, 26, 37, 38).

Effects on Genes

Long before their chemical structure had been determined, the genes had been recognized to be the units of heredity, and their arrangement in linear order on the chromosomes had been established. Within a few years after Muller's discovery, in 1927, that mutations could be induced by x-irradiation (39), mutagenic effects had been measured in a wide variety of organisms, and a theoretical framework had been developed to account for their relationship to

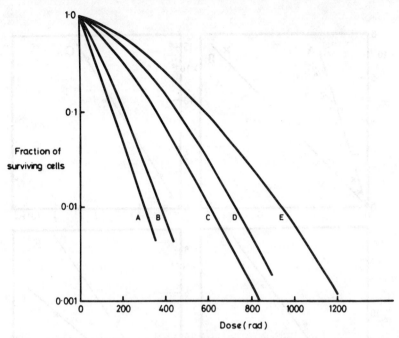

Figure 3 Dose-survival curves for human kidney cells exposed *in vitro* to radiations of differing LET (Reproduced from 35, based on data from 36).

Curve	Radiation	LET (keV/μ)
A	3.4–5.2 MeV α particles	86–140
B	8.3 MeV α particles	60.8
C	26.8 MeV α particles	24.6
D	6.3 MeV deuterons	11.0
E	200 kVp x rays	~2.5

dose, dose rate, and LET (40, 41). Noteworthy is the important extent to which radiation physics and radiation chemistry figured in the development and interpretation of the early dose-response curves (Figure 4; Table 3).

The implication of the data that there might be no threshold for radiation-induced genetic effects—and evidence suggesting that the induction of leukemia might similarly behave as a somatic mutation (42)—led to widespread concern in the 1950's about the long-range health hazards of global radioactive fallout (43, 44, 45), lending impetus to the sentiment for a nuclear test ban treaty.

The interpretation of the mutation process as a single-hit phenomenon did not anticipate the subsequent discovery of DNA repair systems and the dose-rate-dependency of mutagenesis in spermatogonia and oocytes exposed to low-

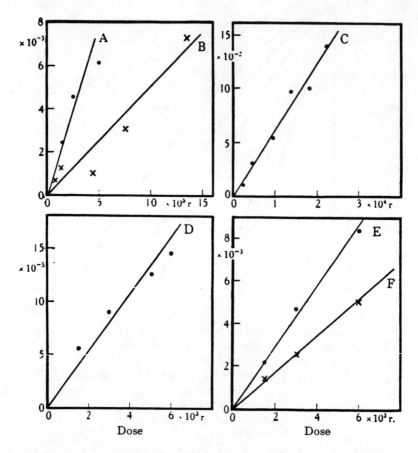

Figure 4 Proportion of visible mutations induced by x-rays as a function of dose. A-B, tobacco mosaic virus (Gowen): A, type to aucuba; B, aucuba to type. C, Neurospora (Demerec et al). D-F, Drosophila melanogaster (Timofeeff-Ressovsky & Delbrück): D, single mutation step (wild type to eosin eye-colour); E, all sex-linked recessive mutations detected by ClB method; F, all sex-linked recessive mutations detected by attached X method. (Reproduced from 40).

LET radiation (46). Although not incompatible with single-hit kinetics for mutagenesis (22, 47), the existence of repair systems in germ cells complicates genetic risk assessment, especially insofar as it involves extrapolation across species differing to unknown degrees in radiosensitivity.

The extent to which newly acquired mutations may affect the prevalence of diseases of complex inheritance is also

Table 3: Early Estimates of the Diameter of the Sensitive Target for Radiation-Induced Mutation in Primordial Germ Cells of Drosophila[a]

Radiation	Hard X-rays or γ-rays	Soft X-rays 2-3A.	Neutrons (Li+D)	α-rays ~3MeV
Relative dose for equal yields of mutations	1.00	1.30	1.45	3.44
Target diameter (in mμ)	—	4.4	9	6.6

[a]From 40.

uncertain at this time. While not strictly a radiobiological problem itself, it nevertheless constitutes a major source of uncertainty in estimates of the risks of radiation-induced genetic detriment to future generations (Table 4).

It is reassuring that extensive study has thus far revealed no evidence of genetic harm in the children of the atomic bomb survivors (48); however, the absence of detectable

Table 4: Estimates of the Heritable Detriment Attributable to Natural Background Ionizing Radiation[a]

Type of Genetic Detriment	Natural Prevalence	Contribution from Natural Background Radiation[b]	
		First Generation	Equilibrium Generations
	(frequency per million live births)		
Dominant traits + x-linked diseases	10,000	75	300
Autosomal + recessive traits	2,500	no increase	45
Chromosomal (structural abnormalities)	400	7	12
Cogenital anomalies	6,000	not estimated	not estimated
Chromosomal (numerical abnormalities)	3,400	not estimated	not estimated
Other multifactorial diseases	650,000	not estimated	not estimated
Total (rounded)	672,300	> 300–900	> 500–3500

[a]From 22
[b]Equivalent to 1 mSv per year, or 30 mSv per parental generation (30 years), average dose to gonads from natural background ionizing radiation.

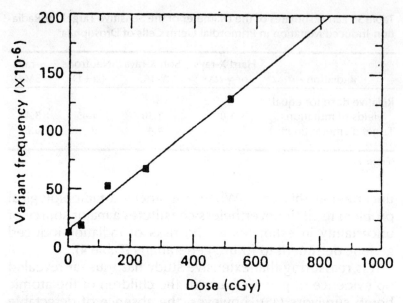

Figure 5 Frequency of hemizygous variations at the glycophorin A locus (presumably somatic mutations) in erythrocytes of a-bomb survivors, in relation to the T65DR dose of a-bomb radiation (From 49).

effects in this population is not astonishing in view of the limited numbers of children it comprises and their relatively small mean parental dose. In view of the dose-dependent increase in the frequency of glycophorin polymorphisms in the erythrocytes of the exposed survivors (Figure 5), which is consistent with expected rates of radiation-induced somatic mutations in their marrow cells, the projected analysis of DNA from the lymphocytes of the survivors and their children may ultimately provide quantitative data on the extent to which the frequency of heritable mutations in the human species is affected by irradiation.

Effects on Chromosomes

Nowhere has radiobiology provided a more precisely quantitative body of dose-effect data than in the study of radiation-induced chromosome aberrations. Thanks to the pioneer studies of Sax (50), Carlson (51), Lea and Catcheside (52), and Giles (53), among others, the relationship of the

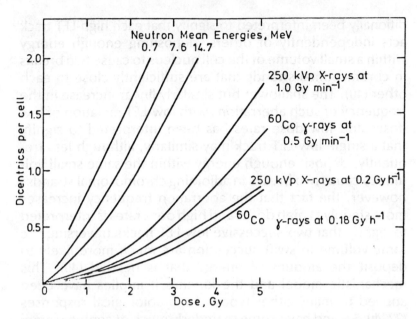

Figure 6 Frequency of dicentric chromosome aberrations in human lymphocytes irradiated *in vitro*, in relation to dose, dose rate, and quality of radiation (From 54).

types and frequencies of chromosome aberrations to the dose, dose rate, and quality of radiation were well documented in nonmammalian cells more than 40 years ago.

In fact, so well has the frequency of chromosome aberrations in human lymphocytes been characterized in relation to the dose (Figure 6) that it now serves as a useful biological dosimeter for assessing the exposure of radiation accident victims (34). As yet, however, the fundamental nature of the underlying molecular lesion—and whether it differs qualitatively with variations in LET—is not clear (38, 55). Furthermore, it has recently been suggested that a small "priming" dose of low-LET radiation may reduce the yield of aberrations resulting from a second dose, owing presumably to the induction of repair (56); this finding implies greater complexity in dose-effect relationships than suspected heretofore.

The linear increase in the frequency of two-break aberrations with increasing dose of high-LET radiation has traditionally been interpreted to signify that each high-LET track acts independently of others, depositing enough energy within a small volume of the cell nucleus to cause two breaks in chromosomal strands that are sufficiently close to each other (40). The shallower but similarly linear increase in the frequency of such aberrations with low-LET radiation of low doses and low dose rates has been interpreted to signify that a single low-LET track may similarly, although less frequently, deposit enough energy within the same small volume to cause two breaks in adjoining chromosomal strands; however, the fact that the aberration frequency increases more steeply at high doses and high dose rates is interpreted to signify that two successive low-LET tracks traversing the same volume in swift succession are much more likely to deposit the amount of energy that is needed (40). This mechanistic model and the kinetics it implies have been applied to many other types of radiobiological responses (22, 40, 57) and have come to underly much of contemporary radiation risk assessment (22, 58, 59); however, the confidence with which the model can be used as a basis for extrapolation into the low dose domain is limited. Most radiobiological effects have been observed only at doses exceeding the range in which cells can be assumed to have been traversed by no more than one track, and the underlying molecular mechanisms are not known precisely for any effect. Estimates of the risks of low doses derived by linear extrapolation on the basis of the model may thus conceivably overestimate or underestimate the actual risks (38).

Carcinogenic Effects

Early in this century, the occurrence of certain cases of skin cancer and leukemia in radiation workers implicated excessive occupational exposure as the causal factor (60, 61, 62), an implication supported by the experimental induction of cancers by irradiation in mice, rats, rabbits, and guinea pigs (Table 5). Systematic study of the carcinogenic effects of radiation has since delineated in some detail how the frequency of carcinogenic effects may vary with the dose,

Table 5: Early Experimental Observations on Radiation Carcinogenesis

Author	Date	Radiation or Isotope	Species	Type of Tumor
Marie et al.	1910	X-ray	Rat	Sarcoma, spindle-celled
Marie et al.	1912	X-ray	Rat	Sarcoma, spindle-celled
Lazarus-Barlow	1918	Ra	Mouse, rat	Carcinoma of skin
Bloch	1923	X-ray	Rabbit	Carcinoma
Bloch	1924	X-ray	Rabbit	Carcinoma
Goebel and Gérard	1925	X-ray	Guinea pig	Sarcoma, polymorphous
Daels	1925	Ra	Mouse, rat	Sarcoma, spindle-celled
Jonkhoff	1927	X-ray	Mouse	Carcinoma-sarcoma
Lacassagne and Vinzent	1929	X-ray	Rabbit	Fibrosarcoma, osteosarcoma rhabdomyosarcoma
Schürch	1930	X-ray	Rabbit	Carcinoma
Daels and Biltris	1931	Ra	Rat	Sarcoma of cranium, kidney, spleen
Daels and Biltris	1931	Ra	Guinea pig	Sarcoma of cranium, kidney, spleen
Daels and Biltris	1937	Ra	Chicken	Carcinoma of biliary tract, osteosarcoma
Schürch and Uehlinger	1931	Ra	Rabbit	Sarcoma of bone, liver, spleen
Uehlinger and Schürch	1931–1947	Ra, Ms-Th	Rabbit	Sarcoma of bone, liver, spleen
Sabin et al.	1932	Ra, Ms-Th	Rabbit	Osteosarcoma
Lacassagne	1933	X-ray	Rabbit	Sarcoma, spindle-celled myxosarcoma
Petrov and Krotkina	1933	Ra	Guinea pig	Carcinoma of biliary tract
Sedginidse	1933	X-ray	Mouse	Carcinoma, spindle-celled
Ludin	1934	X-ray	Rabbit	Chondrosarcoma
Furth	1936	X-ray	Mouse	Leukemia

From 60,61,62.

Table 6: Changes with Time in the Projected Lifetime Excess Mortality from Cancer After 1 Gy Rapid Whole-Body Low-LET Irradiation

Source of Estimate	Additive Risk Projection Model	Multiplicative Risk Projection Model
	(deaths per 10,000 persons)[a]	
BEIR, I, 1972	120	620
UNSCEAR, 1977	100–260	—
BEIR III, 1980	170	500
NUREG, 1985	290	520
UNSCEAR, 1988	400[b]–500[c]	700[c]–1100[b]

[a]Values rounded
[b]Estimate for entire population, based on age-specific risk coefficients.
[c]Estimate for entire population, based on adult age-averaged risk coefficient.

dose-rate, LET, species, sex, age at irradiation, tissue at risk, and other variables (63, 64).

Although the existing data do not suffice to define the carcinogenic risks of low-level irradiation for human populations, they imply that the frequency of tumor-initiating effects may increase as a linear, nonthreshold function of the dose (65) and that the overall lifetime risk per unit dose may be appreciably larger than suspected heretofore (Table 6). It is noteworthy, for example, that the overall death rate from cancer in atomic bomb survivors has increased with their attained age, in parallel with the baseline rate in the general population (Figure 7), a pattern characteristically exhibited by laboratory animals as well (Figure 8). The relationship points to interactions between the carcinogenic effects of radiation and those of age-dependent processes which are consistent with the multicausal, multistage nature of carcinogenesis, but are yet to be explained. This pattern clearly does not apply to cancers of all types; e.g., for leukemia, the excess peaks within the first decade after irradiation, depending on the hematologic type and age at the time of exposure (65, 66). Also, in patients treated with spinal irradiation for ankylosing spondylitis, the total cancer excess appears to be markedly reduced after the second decade following exposure (65).

It is also noteworthy that the extent of life shortening per unit dose is essentially the same in different species of lab-

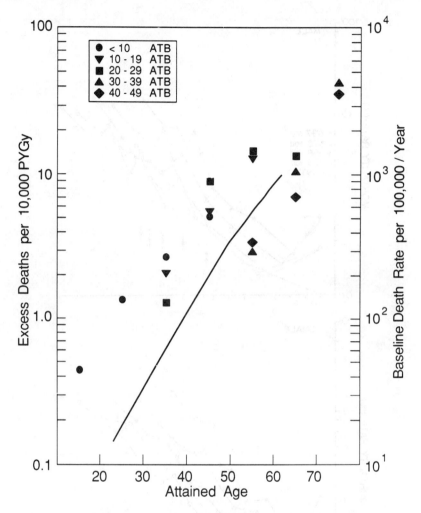

Figure 7 Excess mortality in heavily irradiated (>1 Gy) cohorts of A-bomb survivors of different ages at the time of the bomb (ATB), in relation to attained age. The solid line represents the age-specific baseline cancer death rate for the population of Japan. (A-bomb survivor data from 66; baseline cancer mortality rate data from 67).

oratory animals exposed to whole-body radiation early in life if measured in terms of the percentage increase in their age-specific death rates (Figure 9), in spite of marked differences among the species in the length of the natural life span. To the extent that radiation-induced life shortening is

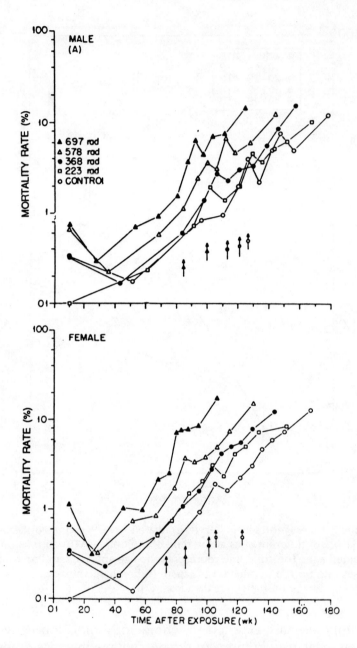

Figure 8 Overall rate of mortality of LAF1 mice in relation to dose and time after acute whole-body gamma irradiation. Arrows indicate mean survival times for 28-day survivors (From 68).

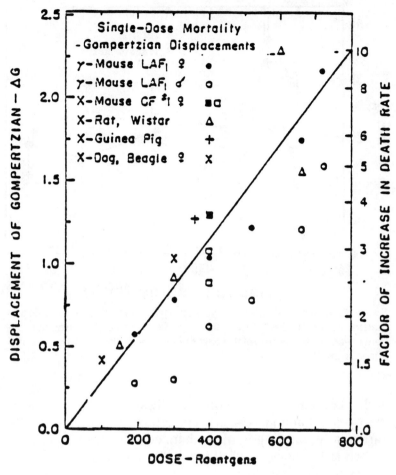

Figure 9 Increase of age-specific death rate in whole-body irradiated animals of different species, as measured by displacement of their Gompertzian life tables (Reproduced from 69, based on data from other sources).

largely attributable to radiation-induced neoplasia (22), the available data for the a-bomb survivors (Figure 7) imply the existence of comparable life shortening effects of whole-body irradiaton in humans.

Still to be fully characterized and explained are large variations in susceptibility to radiation carcinogenesis among different tissues and organs, which bear no apparent relationship to the corresponding baseline cancer rates (70). Also lacking are human data on the influence of dose rate

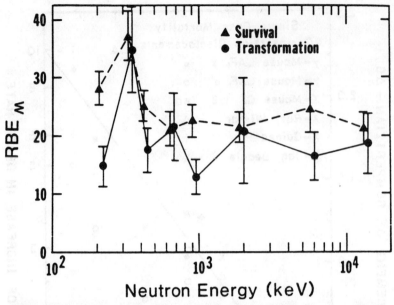

Figure 10 Maximal relative biological effectiveness (RBE$_M$) for survival and transformation of C3H 10T1/2 mouse embryo fibroblasts exposed to monoenergetic neutrons, in relation to the energy of the neutrons (Reproduced from 71).

and of LET, each of which profoundly affects carcinogenesis in laboratory animals (64) and cultured cells (e.g., Figure 10). Particularly puzzling is the enhanced effectiveness of protracted fast neutron irradiation for: 1) carcinogenesis *in vivo*, 2) neoplastic transformation *in vitro* (Figure 11), and 3) mutagenesis in human lymphocytes (73). The marked variations in RBE with the cell type and site of neoplasia also remain to be explained (64, 74). These and other considerations (75) point to the existence of yet to be identified processes that are capable of greatly modifying the evolution of carcinogenic alterations at the cellular and subcellular levels.

Because of limitations in the empirical data, resolution of the remaining questions will require better understanding of the mechanisms of radiation carcinogenesis. Thus, the recent development of refined techniques for investigating the basic mechanisms of neoplasia—through study of the

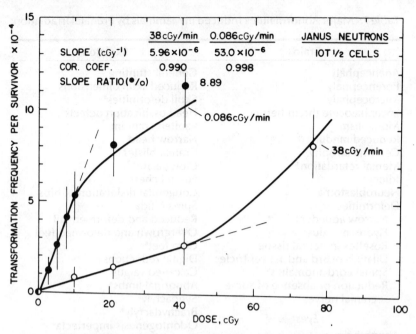

Figure 11 Neoplastic transformation of cultured C3H10T1/2 cells, in relation to the dose and dose rate of neutron irradiation (Reproduced from 72).

roles of oncogenes, tumor-suppressor genes, and other deteminants of cell regulation, for example—holds promise for progress in this field.

Teratogenic Effects

The number of malformations that has been associated with prenatal irradiation of the human embryo, although limited in comparison with the number that has been documented in laboratory animals, is nonetheless considerable (e.g. Table 7). Extensive experimental study of such abnormalities has characterized in detail the types of effects that may occur and the ways in which their frequency and severity may vary in relation to dose, dose rate, LET, and stage in development at the time of exposure (74, 77).

The most noteworthy human findings to date are the dose-dependent increase in frequency of severe mental retarda-

Table 7: Major Abnormalities Induced in Mammals by Prenatal Irradiation

Brain	Skeleton
Anencephaly	General stunting
Porencephaly	Reduced skull dimensions
Microcephaly[a]	Skull deformities[a]
Encephalocele (brain hernia)	Head ossification defects[a]
Mongolism[a]	Vaulted cranium
Reduced medulla	Narrow head
Cerebral atrophy	Cranial blisters
Mental retardation[a]	Cleft palate[a]
Idiocy[a]	Funnel chest
Neuroblastoma	Congenital dislocation of hips
Deformities	Spina bifida
Narrow aqueduct	Reduced and deformed tail
Hydrocephalus[a]	Overgrown and deformed feet
Rosettes in neural tissue	Club feet[a]
Dilation of 3rd and 1st ventricles	Digital reductions
Spinal cord anomalies[a]	Calcaneo valgus
Reduction or absence of some cranial nerves	Abnormal limbs[a]
	Syndactyly[a]
Eyes	Brachydactyly[a]
	Odontogenesis imperfecta[a]
Anophthalmia	Exostosis on proximal tibia Metaphysis
Microphthalmia[a]	
Microcornia[a]	Amelogenesis[a]
Coloboma[a]	Scleratomal necrosis
Deformed iris	
Absence of lens and/or retina	*Miscellaneous*
Open eyelids	Situs inversus
Strabismus[a]	Hydronephrosis
Nystagmus[a]	Hydroureter
Retinoblastoma	Hydrocele
Hypermetropia	Absence of kidney
Congenital glaucoma	Degenerate gonad[a]
Partial albinism	Abnormalities in skin pigmentation
Cataract[a]	
Blindness	Motorial disturbance of extremities
Chorioretinitis[a]	
Ankyloblepharon	Increased probability of Leukemia
	Congenital heart disease
	Deformed ear[a]
	Facial deformities
	Pituitary disturbances
	Dermatomal and myotomal necrosis

From 76.
[a]These anomalies have been found in humans exposed *in utero* to large doses of radiation and have therefore been provisionally attributed to radiation.

tion (74, 78) and the dose-dependent decrease in school achievement test scores (79) in A-bomb survivors who were irradiated prenatally between the 8th and 15th weeks of gestation. The increase in the incidence of severe mental retardation appears to vary as a linear nonthreshold function of the dose and amounts to approximately 40 percent per Sv (74). The fact that it is most marked in those who were irradiated between the 8th and 15th weeks of gestation but was not detectable in those irradiated before the 8th week or after the 26th week is consistent with the chronology of neuroblast proliferation and migration in the developing human brain (74).

Because of their frequency and severity, the developmental abnormalities have implications for radiation risk assessment and radiation protection that call for further study. Their implications also extend to the neurotoxicological effects of other environmental agents—the dose-effect data for lead, for example, suggest that this metal may, similarly, have no threshold for impairment of prenatal brain development (80).

Radionuclide Metabolism

No consideration of the evolution of radiobiology would be complete without acknowledging the wealth of information that has been acquired about the environmental fate and metabolism of radionuclides. To the extent that assessment of the radiobiological implications of radioactivity in the environment is not possible without detailed knowledge of each radionuclide in question, its movement in air, water, and soil, its distribution in the food chain, and its ultimate uptake, distribution, and excretion by man (34, 81, 82), the systematic data that have been gathered on the behavior of radionuclides (83) represent major radiobiological achievements.

The availability of the data—acquired through environmental and ecological studies, as well as through experiments with laboratory animals and other organisms—has enabled assessment of the risks of radionuclides for which no human health effect data are yet available. The data have

Table 8: Early Estimates of Tolerance Dose

Date	Author	1/1000 Erythema Dose (Days)	Calculated (R/day)
1902	Rollins	(film blackening)	10
1925	Mutscheller	3	0.2
1925	Sievert	3	0.2
1926	Solomon	0.3	2.0
1927	Dutch Board	15	0.04
1928	Barclay and Cox	3.5	0.17
1928	Kaye	5	0.12
1931	Adv. Comm. (U.S.)	—	0.2
1932	Failla	30	0.1
1932	Stenstrom	3.7	0.16
1936	Adv. Comm. (U.S.)	—	0.1

From 85.

thus made it possible to develop radiation protection criteria for all radionuclides to which man may be exposed, to minimize the risks and radiobiological consequences of accidental releases of radioactivity, and to place in perspective the risks to populations attributable to irradiation from various natural and man-made sources (34, 84).

Evolution of Risk-Based Standards

Although it became apparent almost immediately after Roentgen's discovery of the x-ray that precautions against excessive exposure were required, scores of pioneer radiation workers were injured before adequate precautions gained wide adoption (4). The earliest efforts to limit exposure to levels that could be tolerated without adverse tissue reaction were hampered by the lack of suitable units of dose and instruments for quantitative dosimetry; nevertheless, they sufficed to provide a basis for significant radiation protection during the first half of this century (Table 8).

Eventually, in the face of evidence implying that there might be no threshold for the genetic effects and carcinogenic effects of radiation (86), the concept of "tolerance" dose gave way to the concept of "maximum permissible" dose, with its corollary principle ALARA (87).

As a result of the abandonment of the threshold concept, the setting of permissible exposure limits has become linked inextricably to risk assessment. Quantitative risk assessment has thus come to play an increasingly important role in radiation protection (30, 84). The demand for more precise estimation of the risks of low-level irradiation has, in turn, called increasingly for research to define the relevant dose-effect relationships.

OUTLOOK FOR THE FUTURE

In spite of the fact that the health effects of low-level irradiation have been investigated more thoroughly than those of any other enviromental agent, important questions persist. As concerns the carcinogenic effects of radiation, for example, it must be recognized that epidemiological studies large enough to measure the effects of doses in the mSv range are not economically or logistically feasible (88). Hence estimation of the carcinogenic risks of low-level irradiation will depend largely on interpolation and extrapolation from observations at higher doses and dose rates, based on assumptions about the relevant dose-incidence relationships and mechanisms of carcinogenesis.

Insofar as human data on the modifying influence of dose rate, LET, and other variables are lacking, extrapolation on the basis of experimental animal data will be indispensable. At the same time, however, since radiation-induced cancers of different types vary in their temporal distributions, more prolonged follow-up of the a-bomb survivors and other irradiated human populations will be necessary to disclose the relationship of risk to age at exposure and attained age. The interactive effects of other risk factors—such as tobacco use, nutritional variables, and inherited susceptibility traits—as well as the particular cellular and subcellular mechanisms (e.g., specific oncogenes and/or tumor-suppressor genes) through which the development of different neoplasms is mediated, also remain to be determined. Thus, progress in assessing the risks of low-level irradiation will call for the coordinated efforts of scientists of many disciplines.

Finally, any consideration of the contributions of radiobiology—past, present, or future—should not overlook the extent to which the concepts and approaches growing out of radiobiological research have contributed to progress in other fields. The earlier advances in radiobiology have derived, in part, from its advantage of having a precise, quantitative system of dosimetry (89). Future progress in radiobiology will depend on the continuing integration of physics, chemistry, and other relevant disciplines into the investigations at hand.

In closing, I should like once again to express my gratitude for having been given this opportunity to pay tribute to Laurie's leadership. I am especially pleased, as we all are, that he and Robena Taylor are both with us today.

The author is grateful to Ms. Lynda Witte for assistance in the preparation of the manuscript.

REFERENCES

1. J. Daniel. "The depilatory action of x-rays," New York Med. Rec. 49,596 (1896).
2. L. G. Stevens. "Injurious effects on the skin," Brit. Med. J. 1,998 (1896).
3. O. Glasser. *The Science of Radiology*. Charles C. Thomas, Springfield, Ill., (1933).
4. P. Brown. *American Martyrs to Science Through the Roentgen Rays*. Charles C. Thomas, Springfield, Ill., (1936).
5. N. Stone-Scott. "X-ray injuries," Am. X-ray J. 1,57–67 (1897).
6. R. S. Stone. "Maximum permissible exposure standards." In *Protection in Diagnostic Radiology*, edited by B. P. Sonnenblick. Rutgers Univ. Press, New Brunswick, NJ, (1959), pp 70–81.
7. H. E. Albers-Schönberg. "Ueber line bisher unbekannte wirkung der Roentgenstrahlen auf den organismus," Munich. Med. Wochschr. 50,1859 (1903).
8. A. von Frieben. "Hodenverauderungen bei tieren nach Rontgenstrahlung," Munchen Med. Wuhnschi 50,2295 (1903).
9. J. Bergonié and L. Tribondeau. "Action des rayons X sur le testicule du rat blanc," Compt. rend Soc. biol. 57,400 (1904).
10. J. Bergonié and L. Tribondeau. "Action des rayons X sur le spermatozoides de l'homme," Compt. rend Soc. biol. 57,595 (1905).
11. J. Bergonié and L. Tribondeau. "Action des rayons X sur le testicule de rat blanc," Compt. rend Soc. biol. 58,154, 678 and 1029 (1906).

12. J. Bergonié and L. Tribondeau. "Interprétation de quelques résultats de la radiotherapie et assai de fixation d'une technique rationale," C.R. Acad. Sci. 143,983–985 (1906).
13. C. Regaud and J. Blanc. "Actions des rayons X sur les diverses générations de lignee spermatique; extréme sensibilité de spermatogonies aux rayons," Compt. rend Sco. biol. 61,163 (1906).
14. M. Seldin. Uber die Wirkung der Rontgen- und Radiumstrahlen auf innere Organne und Gesamtorganismus der Tiere," Fortschr. Geb. Rontgenstrahlen 7,322 (1904).
15. F. T. Brown and A. T. Osgood. "X-rays and sterlility," Am J. Surg. 18,179–182 (1905).
16. F. Villemin. "Rayons X et activite genitale," Compt. rend. Acad. Ac. 142,723 (1906).
17. C. Regaud. "Le rythme alternant de multiplication cellulaire et la radiosensitivité du testicle," Compt. rend. Soc. Biol. 87,427 (1922).
18. R. Ferroux and C. Regaud. "Est-il possible de steriliser le testicule du lapin adulte par une dose de rayons X sans produire le lesion grave de la peau?" Compt. rend. Soc. Biol. 97,330 (1927).
19. E. F. Oakberg and E. Clark. "Species comparisons of radiation response of the gonads." *Proceedings of the International Symposium on the Effects of Ionizing Radiation on the Reproductive System.* (1964), pp. 11–24.
20. C. G. Heller. Data cited in: *Radiobiological Factors in Manned Space Flight,* edited by W. H. Langham. National Academy of Sciences, Washington, D.C., (1967), pp. 127–128.
21. E. F. Oakberg. "A new concept of spermatogonial stem-cell renewal in the mouse and its relationship to genetic effects," Mutation Res. 11,1–7 (1971).
22. United Nations Scientific Committee on the Effects of Atomic Radiation. *Ionizing Radiation: Sources and Biological Effects,* Report to the General Assembly, with Annexes. United Nations, New York (1982).
23. M. Strandqvist. "Studien über die kumulative Wirkung der Röntgenstrahlen bei Fraktionierung," Acta Radiol. 55, 1, 1944.
24. P. Rubin and G. W. Casarett. *Clinical Radiation Pathology,* Vol. I and II. W. Saunders, Philadelphia, (1968).
25. R. S. Stone. "The concept of a maximum permissible exposure," Radiology 58,639 (1952).
26. International Commission on Radiological Protection. *Nonstochastic Effects of Radiation.* ICRP Publication No. 41. Pergamon Press, Oxford (1984).
27. M. W. Charles. "The biological bases of radiological protection criteria for superficial, low penetrating radiation exposure," Radiation Protection Dosimetry 14,79–90 (1986).
28. F. F. Hahn, R. O. McClellan, B. B. Boecker, and B. A. Muggenburg. "Future development of biological understanding of radiation protection: implications of nonstochastic effects," Health Phys. 55,303–315 (1988).

29. International Commission on Radiological Protection. "Recommendations of the International Commission on Radiological Protections (as adopted September 17, 1965)." In: *ICRP Publication No. 9.* Pergamon Press, Oxford (1966).
30. International Commission on Radiological Protection. "Recommendations of the International Commission on Radiological Protection." In: *ICRP Publication 26.* Annals of the ICRP 1, No. 3. Pergamon Press, Oxford (1977).
31. T. T. Puck and P. I. Marcus "Action of x-rays on mammalian cells," J. Exper. Med. 103,653–66, 1956.
32. M. M. Elkind and H. Sutton. "Radiation response of mammalian cells grown in culture. 1. Repair of x-ray damage in surviving Chinese hamster cells," Radiat. Res. 13,556–593 (1960).
33. V. P. Bond, T. M. Fliedner, and J. O. Archambeau. *Mammalian Radiation Lethality: A Disturbance in Cellular Kinetics.* Academic Press, New York (1965).
34. United Nations Scientific Committee on the Effects of Atomic Radiation. *Sources, Effects, and Risks of Ionizing Radiation.* Report to the General Assembly, with Annexes. United Nations, New York (1988).
35. S. B. Field. "An historical survey of radiobiology and radiotherapy with fast neutrons," Curr. Top. Radiat. Res. 11,1–36 (1976).
36. G. W. Barendsen. "Responses of cultured cells, tumors and normal tissues to radiations of different linear energy transfer 1968." In *Current Topics in Radiation Research,* Vol. 4. Ebert and Howard, Amsterdam, (1968), pp. 293–356.
37. W. K. Sinclair. "Failla Memorial Lecture: Risk, research, and radiation protection," Radiat. Res. 112,191–216. (1987).
38. D. T. Goodhead. "Spatial and temporal distribution of energy," Health Phys. 55,231–240 (1988).
39. H. Muller, Jr. "Artificial transmutation of the gene," Science 46,84–87 (1927).
40. D. E. Lea. *Actions of Radiation on Living Cells.* MacMillan, New York (1947).
41. J. H. Muller. "The manner of production of mutations by radiation." In *Radiation Biology, Vol. 1: High Energy Radiation,* edited by A. Hollaender. McGraw-Hill, New York, (1954), pp. 475–626.
42. E. B. Lewis. "Leukemia and ionizing radiation," Science 125,965–975 (1957).
43. National Academy of Sciences. Committee on the Biological Effects of Atomic Radiation. *The Biological Effects of Atomic Radiation.* National Academy of Sciences, Washington, D.C. (1956).
44. United Nations Scientific Committee on the Effects of Atomic Radiation. *Report to the General Assembly,* (Official Records: Thirteenth Session, Supplement No. 17). United Nations, New York (1958).
45. United Nations Scientific Committee on the Effects of Atomic Radiation. *Report to the General Assembly,* (Official Records: Seventeenth Session, Supplement No. 16). United Nations, New York (1962).

46. W. L. Russell. "Recent studies on the genetic effects of radiation in mice," Pediatrics (Suppl) 41,223–230 (1968).
47. United Nations Scientific Committee on the Effects of Atomic Radiation. *Ionizing Radiation: Levels and Effects*. (Official Records: Twenty-seventh Session, Supplement No. 25). United Nations, New York (1972).
48. J. V. Neel, et al. "Search for mutations altering protein charge and/or function in children of atomic bomb survivors: final report," Am. J. Hum. Genet. 42,663–676 (1988).
49. R. G. Langlois, W. L. Bigbee, S. Kyoizumi, N. Nakamura, M. A. Bean, M. Akiyama, and R. H. Jensen. "Evidence for increased somatic cell mutations at the glycophorin A locus in atomic bomb survivors," Science 236,445–448, 1987.
50. K. Sax. "Types and frequencies of chromosomal aberrations induced by x-rays," Cold Spring Harbor Symposium 9,93 (1941).
51. J. G. Carlson. "Effects of x-rays on grasshopper chromosomes," Cold Spring Harbor Symposium 9,104 (1941).
52. D. E. Lea and D. G. Catcheside. "Induction by radiation of chromosome aberrations in tradescantia," J. Genet 44,216 (1942).
53. N. H. Giles. "Comparative studies of the cytogenetic effects of neutrons and x-rays," Genetics 28,398 (1943).
54. D. C. Lloyd and R. J. Purrott. "Chromosome aberration analysis in radiological protection dosimetry," Radiation Protection Dosimetry 1,19–28 (1981).
55. I. R. Radford, G. S. Hodgson, and J. P. Mathews. "Critical DNA target size model of ionizing radiation-induced mammalian cell death," Int. J. Radiat. Biol. 54,63–79 (1988).
56. S. Wolff, V. Afzal, J.K. Wiencke, G. Olivieri, and A. Michaeli. "Human lymphocytes exposed to low doses of ionizing radiations become refractory to high doses of radiation as well as to chemical mutagens that induce double-strand breaks in DNA," Int. J. Rad. Biol. 53,39–48 (1988).
57. A.M. Kellerer and H.H. Rossi. "The theory of dual radiation action," Curr. Top. Radiat. Res. 8,85–158 (1972).
58. A.C. Upton. "Radiobiological effects of low doses: implications for radiological protection," Radiat. Res. 71,51–74 (1977).
59. National Academy of Sciences. Committee on the Biological Effects of Ionizing Radiation. *The Effects on Population of Exposure to Low Levels of Ionizing Radiation*. National Academy of Sciences, Washington, D.C. (1980).
60. A. Lacassagne. "Les cancers produits par les rayonnements electromagnetiques." In: *Actualities Scientifiques et Industrielles, No. 975*. Herman & Cie, Paris (1945).
61. A. Lacassagne. "Les cancers produits par les rayonnements corpusculaires; mechanisme presumable de la cancerisation par les rayons." In: *Actualites Scientifiques et Industrielles, No. 981*. Hermann & Cie, Paris (1945).

62. J. Furth and E. Lorenz. "Carcinogenesis by ionizing radiations." In *Radiation Biology, Vol. 1,* edited by A. Hollaender. McGraw-Hill, New York, (1954), pp. 1145–1201.
63. A.C. Upton, R.E. Albert, F. Burns, and R.E. Shore, Editors. *Radiation Carcinogenesis.* Elsevier, New York (1986).
64. R.J.M. Fry and J.B. Storer. "External radiation carcinogenesis," Adv. in Radiat. Biol. 13,31–90 (1987).
65. A.C. Upton. "Cancer induction and nonstochastic effects," Brit. J. Radiol. 60,1–16 (1987).
66. Y. Shimizu, H. Kato, and W.J. Schull. *Life Span Study Report 11, Part 2. Cancer Mortality in the Years 1950–1985 Based on the Recently Revised Doses (DS86). Technical Report RERF TR 5-88.* Radiation Effects Research Foundation, Hiroshima (1988).
67. T. Hirayama, Editor. *Comparative Epidemiology of Cancer in the U.S. and Japan. Mortality.* U.S.-Japan Cooperative Cancer Research Program, Tokyo (1977).
68. A.C. Upton, A.W. Kimball, J. Furth, K.W. Christenberry, and W.H. Benedict. "Some delayed effects of atom-bomb radiations in mice," Cancer Res. 20,1–62 (1960).
69. G.A. Sacher. "The Gompertz transformation in the study of the injury-mortality relationship: application to late radiation effects and aging." In: *Radiation & Aging,* edited by P.J. Lindop and G.A. Sacher. Taylor and Francis, London, (1966), pp. 411–416.
70. G.W. Beebe. "A methodologic assessment of radiation epidemiology studies," Health Physics 46 No. 4,745–762 (1984).
71. R.C. Miller, C.R. Geard, D.J. Brenner, K. Komatsu, S.A. Marino, and E.J. Hall. "Neutron-energy-dependent oncogenic transformation of C3H 10T12 mouse cells," Radiation Research 117,114–127 (1989).
72. C.K. Hill, A. Han, and M.M. Elkind. "Fission-spectrum neutrons at low dose rate enhance neoplastic transformation in the linear, low dose region (0–10 cGy)," Int. J. Radiat. Biol. 46,11–15 (1984).
73. A.C. Upton. "Unresolved questions on the risks from high-LET radiation." In: Proceedings of the Workshop on Risks from Radium and Thorotrast, Bethesda, Maryland, October 3–5, 1988. In presss (1989).
74. United Nations Scientific Committee on the Effects of Atomic Radiation. *Genetic and Somatic Effects of Ionizing Radiation.* Report to the General Assembly, with Annexes. (Forty-first Session, Supplement No. 16). United Nations, New York (1986).
75. A.R. Kennedy. The conditions for the modification of radiation transformation *in vitro* by a tumor promoter and protease inhibitors," Carcinogenesis 6,1441–1445 (1985).
76. A.B. Brill and E.H. Forgotson. "Radiation and congenital malformations," Am. J. Obstet. Gynecol. 90,1149–1168 (1964).
77. United Nations Scientific Committee on the Effects of Atomic Radiation. *Sources and Effects of Ionizing Radiation.* Report to the General Assembly, with Annexes. (Thirty-second Session, Supplement No. 40). United Nations, New York (1977).

78. M. Otake and W. Schull. "*In utero* exposure to a-bomb radiation and mental retardation: a reassessment," Brit. J. Radiol 57,409–414 (1984).
79. W.J. Schull, M. Otake, and H. Yoshimara. *Effect on Intelligence Test Score of Prenatal Exposure to Ionizing Radiation in Hiroshima and Nagasaki: A Comparison of the T65DR and DS86 Dosimetry Systems.* (Technical Report RERT TR3-88). Radiation Effects Research Foundation, Hiroshima (1988).
80. D. Bellinger, A. Leviton, C. Watermaux, H. Needleman, and M. Rabinowitz. Longitudinal analyses of prenatal and postnatal lead exposure and early cognitive development," New Eng. J. Med. 316,1037–1043 (1987).
81. National Council on Radiation Protection and Measurements. *Radiological Assessment: Predicting the Transport, Bioaccumulation and Uptake by Man of Radionuclides Released to the Environment.* (NCRP Report No. 76). National Council on Radiation Protection and Measurements, Bethesda, MD (1985).
82. J.E. Till. "Modeling the outdoor environment—new perspectives and challenges," Health Physics 55,331–338, 1988.
83. J.N. Stannard. *Radioactivity and Health: A History,* edited by R.W. Baalman, Jr. DOE/RL/01830-T59 (DE88013791). NTIS, Springfield, VA (1988).
84. National Council on Radiation Protection and Measurements. *Recommendations on Limits for Exposure to Ionizing Radiation.* (NCRP Report No. 91). National Council on Radiation Protection and Measurements, Bethesda, MD (1987).
85. K.Z. Morgan. "History of damage and protection from ionizing radiation." In *Principles of Radiation Protection: A Textbook of Health Physics,* edited by K.Z. Morgan and J.E. Turner. Wiley, New York, (1967), pp. 1–75.
86. International Commission on Radiological Protection. "International recommendations on radiological protection 1950," Brit. J. Radiol 24,46–53 (1951).
87. L.S. Taylor. *Organization for Radiation Protection—The Operation of the ICRP and NCRP (1928–1974).* DOE/TIC-10124. NTIS, Springfield, VA (1980).
88. C.E. Land. "Estimating cancer risks from low doses of ionizing radiation," Science 290,1197–1203 (1980).
89. A.C. Upton. "Evolving perspectives on the concept of dose in radiobiology and radiation protection," Health Phys. 55,605–614 (1988).

Warren K. Sinclair, President of the NCRP presenting commemorative plaque to the lecturer.

Scientific Session

Nonionizing Radiation

S. James Adelstein
Chairman

Protection Against Nonionizing Electromagnetic Radiation, an Evolutionary Process

G.M. Wilkening
AT&T Bell Laboratories
Murray Hill, NJ

It is an interesting historical fact that within the last 60 years the National Council on Radiation Protection and Measurements (NCRP) has devoted the overwhelming majority of its efforts to the assessment and control of the effects of ionizing radiation. The first nonionizing radiation project commissioned by the NCRP Board occurred in the early part of this decade with the emergence of Scientific Committee 39 on Microwaves and its Report No. 67 Radiofrequency Electromagnetic Fields: Properties, Quantities and Units, Biophysical Interaction, and Measurements (1).

The explanation for this difference in program emphasis within NCRP, derives from an understanding of the differences in interaction mechanisms between ionizing and nonionizing radiation. Ionizing radiations contain considerably greater quantum energy, sufficient to dislodge electrons from atoms to form ionized species whose cumulative harmful effects are potentially much more hazardous than the more immediate effects associated with the longer nonionizing wavelengths.

To the average layman the term nonionizing radiation is new and unfamiliar. The practitioners in the radiological health sciences, such as those in the audience, still tend to specialize either in ionizing radiation or nonionizing radiation, despite some recent progress toward more open dialogue. Radiofrequency and microwave engineers have gone to a lot of trouble to replace the word "radiation" with RF or microwave "energy" so as to avoid any possible confusion with ionizing radiation and its perceived dangers. Engineers associated with 60 Hz power transmission systems feel completely justified in doing so since most exposures occur in static fields.

I have tried over the past two or three decades to look for ways to encourage the belief that we are not in two totally separate professional fields—that our common interest is to protect people from excessive levels of radiation of all types. My intention today was to look back briefly over the past 60 years to recollect a few key events that helped to improve our understanding and control of the nonionizing radiations. However, in an effort to bring our two disciplines a little closer, I have decided to go back in time well beyond a mere 60 years.

I have decided to call to your attention the creation of the universe, with due credits to Michael Turner of Fermilab (2). (Figure 1) This is the hot big bang model, the standard model of cosmology, that purports to catalogue events from about 10^{-48} seconds after the big bang to this very moment in time, some 15 billion years later, and all on the same scale. Since all of this is speculative, it is nothing short of amazing that there seems to be a consensus among most astrophysicists that the depiction given in Figure 1 is "reasonable." In the midst of all the cosmic events cited here I wish to draw your attention to the presence of microwave radiation. The microwave background is our oldest fossil. It was discovered in 1965 by Penzias and Wilson (3) and is considered to be relic radiation from a hot dense phase of the universe. This residual "warmth" from the primeval fireball, predicted for some time prior to its discovery, appears to follow the emissivity curve for a black body at $2.75 K \pm 0.05 K$. The simultaneous presence of ionizing and nonionizing events long before and long after the emergence of life on our planet should remind us that we all coexist in an electromagnetic continuum in spite of any effort to treat portions of the spectrum as separate subjects.

The diffuse photon spectrum of the universe is shown in Figure 2 (4). In this case, gravitational waves are thought to have been generated by quantum fluctuations during an inflationary epoch. The secondary background radiations arose once these fluctuations became nonlinear and collapsed. The far-infrared background is thought to have originated from dust emissions, whereas the near infrared, optical and UV backgrounds are thought to have arisen from stellar and quasar emissions.

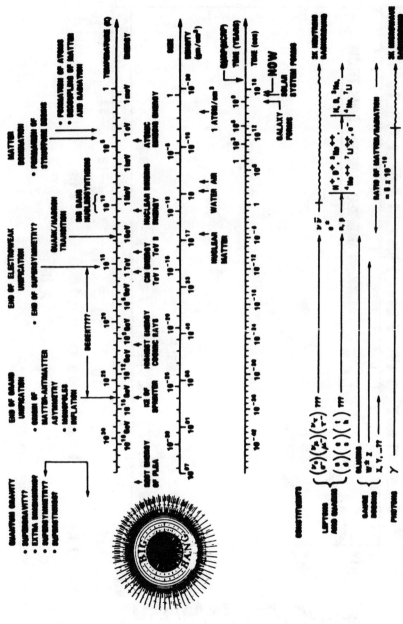

Fig. 1. The Complete History of the Universe (2)

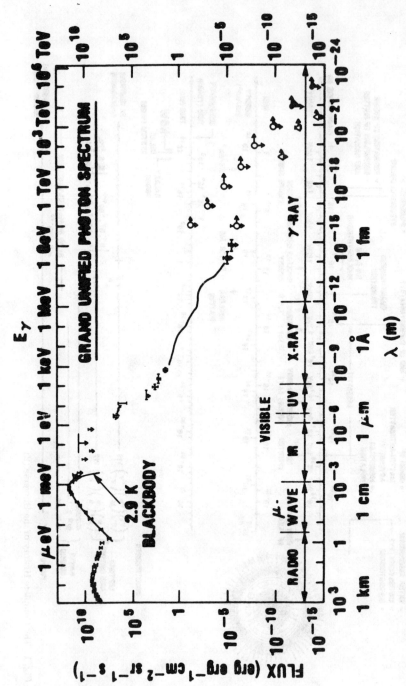

Fig. 2. The Diffuse Photon Spectrum of the Universe (4)

The x rays presumably originated from quasar emissions and cluster collapse. Even though the nonionizing radiations have a lower quantum energy than ionizing radiations the fact that they appear to have a higher flux is of interest in terms of our present knowledge of health effects.

With the formation of our solar system, the natural background of nonionizing electromagnetic radiation in the terrestrial biosphere arises primarily from our sun and from reradiation from the earth. The maximum solar spectral irradiance (E_λ) is found at a wavelength of 0.5084 μm; the maximum for terrestrial black body reradiation is found at 10.06 μm. For all practical purposes the RF/microwave levels of any significance at the earth's surface arise from man-made sources. Also noteworthy is the fact that cosmic radiation and natural terrestrial radioactivity can create nonionizing electric fields of the order of 100 to 200 V/m, in the earth's atmosphere, and the presence of lightning may produce fields as high as 10^3 V/m in the ELF region of 3 to 300 Hz and in the VLF region of 1 to 10 kHz.

Since this annual meeting celebrates 60 years of NCRP activity, permit me to mention briefly a few historical events plus some nonionizing radiation projects that have been initiated and completed by the NCRP. There are many such projects under development by NCRP committees at this time. These include the biological effects of magnetic fields, the biological effects of extremely low frequency (ELF) radiation, the formation of a new committee on ultraviolet radiation and a soon to be published "Practical Guide on the Evaluation of Human Exposures to Radiofrequency Radiation." Because ultrasound is not considered electromagnetic energy, that subject will be covered later today in a separate presentation. And, since the only reports on nonionizing electromagnetic radiations published by NCRP are those devoted to RF radiations, my plan is to begin with a brief chronology of events leading to the NCRP radiofrequency exposure criteria. (Table 1)

Prior to the time NCRP Report 67 was published in 1981 there was very little standardization of quantities and units. Most investigators were tolerating the use of terms like "dose," "dose rate," "absorbed power density" in the very low frequency ranges. The System Internationale (SI) was being ignored. I believe the proposed use of the term "absorbed power density" produced sufficient revulsion in the rank and file that everyone agreed that the quantities and units had to be improved. The use of such loose terminology reflected poorly on the quality of work being performed and it caused difficulty when trying to compare the results of different studies. Report 67 was a key forerunner in the NCRP entry into the nonionizing radiation arena in that it laid down meaningful, consistent definitions after an exhaustive review and reso-

TABLE 1—*Chronology of Selected Events Associated with RF Exposure Limits in the U.S.A.*

Prior to 1950	Fragmented reports on electrical excitation of tissue.
1952	Dr. Fred Hirsch reported eye damage in the AMA Archives of Industrial Hygiene and Occupational Medicine (5). Estimated "power density" to which man was exposed: 100 mW/cm^2.
1953	Naval Medical Research Meeting; concluded that 100 mW/cm^2 was detrimental. Schwan estimated 10 mW/cm^2 as possible exposure criterion. (6) Bell Laboratories estimated safe level at 0.1 mW/cm^2 (7). (Private speculation that Russians may have miscopied this value as 0.01 mW/cm^2).
1955	General Electric established a level of 1 mW/cm^2 (8). Although many power density values were proposed there was little scientific basis for them because of the paucity of definitive biological data. Mayo Clinic attendees reviewed data on RF-induced cataracts in animals (9). Schwan and Li proposed 10 mW/cm^2 at all RF frequencies (6). Basis: metabolism of food generates energy at a rate of 5 mW/cm^2; doubling this should not cause harm. No reported detrimental effects at levels of approximately 10 mW/cm^2 (10). The need for additional biological substantiation was recommended by Schwan.
1956	Microwave "hearing" reported (RADAR) (11).
1957–58	The 10 mW/cm^2 value adopted by Bell Laboratories, GE, Army, Navy, Air Force (12). Consensus was thermal basis although nonthermal possibility discussed.
1957–61	Tri-Service Conferences. Col. G. M. Knauf, U.S.A.A.F. reported international concurrence with 10 mW/cm^2 value (13).
1961	Soviet standard 0.01 mW/cm^2 maximum in one day (based on reported low level CNS effects). Soviet regulations required eye protection at 1 mW/cm^2 (14). Translations of Soviet work often inaccurate. Also few technical details available on experimental design, measurements, and statistical treatment.
1962	Documentation of low intensity microwave irradiation of the US Embassy in Moscow (15). Security issues.
1966	ANSI C95 Radiofrequency Standard adopted (frequency independent from 10 MHz to 100 GHz) (16). Johns Hopkins Applied Physics Laboratory attempted modeling of U.S. exposure criteria.
1968	Senate Commerce Committee Meetings (17).
1968	Radiation Control for Health and Safety Act (PL 90-602) Technical Electronic Products Radiation Safety Standards Committee (TEPRSSC) established (18).
1969	Bureau of Radiological Health (BRH) standards adopted under Public Law 90-602: (19) TV, microwave ovens, electron tubes, diagnostic x ray equipment. Symposium on Biological Effects and Health Implications of Microwave Radiation, Richmond, Virginia. Renewed interest in research. First Symposium since the Tri-Service meetings (20).

1970	Hirsch reevaluated 1952 report (21). Claimed exposure level was much greater than 100 mW/cm^2.
1971	Electromagnetic Radiation Management Advisory Council (ERMAC) established in the President's Office of Telecommunications Policy (OTC). Five year program, multi-agency, $63 million proposed budget. One objective was the determination of the long term effects of the low level microwave radiation.
1973	Senate Committee on Commerce. Hearings on Radiation Control for Health and Safety Act of 1968. (Public Law 90-602) (22). Consumer Union recommends against purchase of microwave ovens, claiming unknown biological effects (23). TEPRSSC recommends the 1 and 5 mW/cm^2 microwave oven emission limits to BRH. American Home Appliances Manufacturers (AHAM) make presentations in support of the manufacture of microwave ovens (23).
1974	National Academy of Science Committee on the Biosphere Effects of Extremely Low Frequency Radiation (ELF) (24). SEAFARER, formerly SANGUINE project.
1974	ANSI C95 1966 Standard reconfirmed with minor revisions (25).
1975	USAF proposes 50 mW/cm^2 (10 kHz–10 MHz) and 10 mW/cm^2 (10 MHz–300 GHz) (26). Early U.S. evidence of a frequency-dependent exposure criterion.
1976	Irradiation of US Embassy continues in Moscow (27). Lilienfeld epidemiological study (28).
1981	NCRP Report No. 67—Radiofrequency Electromagnetic Fields: Properties, Quantities and Units, Biophysical Interaction and Measurements. The report develops and proposes the quantity: specific absorption rate (SAR) (1).
1981	WHO Classifies East European Standards: Group I (USSR): 10μW/cm^2 not to exceed 1 mW/cm^2; Group II: (GDR, Poland, Czechoslovakia): general population standard 10μW/cm^2–100μW/cm^2 (29).
1982	New ANSI C95 Standard. Frequency dependent and based on a whole-body averaged specific absorption rate of 0.4 W/kg (30).
1986	NCRP Report No. 86 Biological Effects and Exposure Criteria for Radiofrequency Electromagnetic Fields (31).

lution of the many disparate definitions found in electrical engineering dictionaries. Report 67 covered such fundamentals as field quantities, electrostatics, magnetostatics, electrodynamics, analogies with circuit quantities, wave propagation, polarization, modulation, reflection/refraction, standing waves, spherical waves, transmission line fields, antennas, energy transfer processes and field and SAR measurements. The appendices contained excellent information on the natural background of nonionizing radiation in the biosphere; RF and microwave absorption in biopolymers, and molecular dynamics in the presence of electromagnetic field pertubation, to name just a few. A major contribution of this report was the introduction of the term specific absorption rate (SAR) into the RF/microwave radiation vocabulary.

SAR was defined as the rate at which electromagnetic energy is absorbed at a point in a medium, per unit mass of the medium, expressed in W/kg. Since energy absorption is a continuous and differentiable function of space and time one may speak of its gradient and rate. Therefore the time derivative of the incremental energy (dW) absorbed in an incremental mass (dm) contained in a volume element (dV) of a given density (ρ) is expressed as:

$$SAR = \frac{d}{dt}\left(\frac{dW}{dm}\right) = \frac{d}{dt}\left(\frac{dW}{\rho dV}\right)$$

The term SAR should not be confused with a measure of the rate of heating of a tissue although measuring the rate of temperature rise is one of many methods for measuring SAR. SAR is strictly a dosimetric quantity that provides no information as to interaction mechanism, *i.e.*, thermal or athermal.

The culmination of the many investigations over a period of some 40 years was the creation in 1982 of the ANSI C95 Radiofrequency Protection Guide (RFPG) (30) using basic concepts and definitions developed in NCRP Report No. 67. For the first time in the development of RFPGs, due cognizance was taken of such factors as body size, mass and orientation, the polarization of the incident wave, the frequency and intensity of the radiation, the presence of reflective surfaces, and whether the irradiated subject was in conductive contact with a ground plane.

Because of sparse quantitative human data on which to model an exposure criterion, Subcommittee IV of the ANSI C95 Committee conducted a review of animal exposure data to determine the most significant, reliable, and independently replicated biological/physiological endpoints produced at the lowest specific absorption rate. Thresholds of behavioral impairment (behavioral disruption) were found within a narrow range of whole-body averaged SARs, ranging from 4 to 8 W/kg. (31) The corresponding range of power densities was 8 to 140 mW/cm². Since the thresholds of behavioral disruption in primates were found to be approximately 4 W/kg this value was chosen as the working threshold for untoward effects in humans in the frequency range of 3 MHz to 100 GHz. (At frequencies below 3 MHz body surface interactions and electric shock are the relevant criteria, not SAR.) The safety margin is a factor of 10 applied to a SAR value of 4 W/kg, hence compliance with the values in the tabular representation of the ANSI C95.1-1982 RFPG shown in Table 2 will ensure that no person will be exposed to a specific absorption rate in excess of 0.4 W/kg, averaged over any 6 minute interval in the frequency range of 3 MHz to 100 GHz. Although the frequency range

TABLE 2—*Exposure Criteria and Rationale*

Frequency (f) Range	Equivalent Power Density	(Electric Field)^2	(Magnetic Field)^2
MHz	mW/cm^2	V^2/m^2	A^2/m^2
0.3–3	100	4×10^5	2.5
3.0–30	900/f^2	$4 \times 10^3(900/f^2)$	0.025(900/f^2)
30–300	1.0	4×10^3	0.025
300–1500	f/300	$4 \times 10^3(f/300)$	0.025(f/300)
1500–100,000	5.0	2×10^4	0.125

of primary interest with respect to SAR is approximately 3 to 3000 MHz, the so called body resonance range, it is believed that limiting the whole-body averaged SAR to 0.4 W/kg will automatically provide an additional safety factor for exposure to frequencies above this range.

The main difference between the ANSI criteria (30) and those proposed by NCRP Scientific Committee 53 is the inclusion of a separate more stringent general population exposure criterion in the NCRP document. (31) See Table 3. ANSI proposed that its criteria apply to persons occupationally exposed in the workplace as well as persons in the general population on the basis that RF exposures produce threshold effects and that a proper exposure envelope had been included to protect all persons. NCRP countered with the rationale that occupationally exposed persons are usually well informed about potential hazards, and are free to determine which risks are acceptable to them, but persons in the general population are generally unaware of RF exposure risks, have little control over exposure levels, and they represent a much larger population comprised of people who are possibly more vulnerable to exposure *e.g.*, the aged, pregnant females, infants, children and the chronically ill. Scientific Committee 53 (31) has recommended that the exposure criterion for the general public be set at a level equal to one-fifth that for occupational groups. That is, the whole body averaged SAR for the general population should not exceed 0.08 W/kg with an averaging period of 30 minutes. This reduction is based on the relative exposure periods of the two groups: 168 hours vs. 40 hours per week, *i.e.*, 40/168 equals approximately 0.2.

Since these data have already been reviewed with this audience on a previous occasion and since we have traversed historical events and are now dealing with current issues in protecting people exposed to RF radiation, I will conclude with a few observations about these latest exposure criteria and our future direction:

TABLE 3—*Radiofrequency Protection Guide Rationale: ANSI vs. NCRP*

Parameter	ANSI	NCRP
Recognition of Whole-Body Resonance	yes	yes
Incorporation of Dosimetry (SAR)	yes	yes
Database	Relatively Small (32 Citations)	Large
Most Significant Biological Endpoint	—Behavioral Disruption—	
Whole-Body Averaged SAR Associated with Behavioral Disruption	4–8 W/kg	3–9 W/kg
Limiting Whole-Body SAR	0.4 W/kg	0.4 W/kg 0.08 W/kg*
Averaging Time	0.1 Hour	0.1 Hour 0.5 Hour*
Criterion for Limits Below 3 MHz	—Surface Effects (E)— (Perception, Electric Shock)	
Criterion for Localized Exposure	—Whole-Body Averaged SAR < 0.4 W/kg— Peak SAR < 8 W/kg	
Special Criterion for Modulated Fields	no	yes (For Occupational Exposure)

*General Population

1. The limits of exposure below 30 MHz apply to free space conditions where a person is not in contact with any object, including the ground. For other conditions, such as standing on the ground with insulated shoes and making contact with a grounded object, or being grounded and touching an insulated metal object, the limits must be lowered, and a case-by-case evaluation must be made. The standard does not tell you exactly how to do this, other than to determine the applicable exposure limits through the use of three criteria: whole-body SAR (0.4 W/kg), maximal local SAR (8 W/kg), and RF burns at the point of contact (200 mA). A great deal of caution is in order in applying the exposure criterion to situations where persons are exposed to high fields at frequencies below 3 MHz. Also, at higher frequencies the local, or spatial peak SAR can be incredibly high while still conforming to the whole-body averaged value of 0.4 W/kg.
2. Although the whole-body average SAR may not exceed a value of 0.4 W/kg, data obtained with simulated human models, indicate that the localized SAR values ("hot spots") may reach 20 times this value. Since this is also assumed to be the case for the animal species in the studies used to develop the RFPG, it was concluded that 8 W/kg in any gram of tissue is acceptable as long as the whole body SAR does not exceed 0.4 W/kg averaged over 0.1 hour (the occupational limit). We also need more biological underpinning of the data before we can feel comfortable about permitting such high localized SARs. We need this information in order to make a more quantitative assessment of the following apparently conflicting statements contained in NCRP Report No 86. (31) The first statement is: "In those cases where there are highly intense focal concentrations of RFEM energy ("hot spots"), this knowledge should supersede whole body values and lead to a corresponding reduction in permissible exposure level." The second statement is: "The deposition of EM energy in all or certain parts of the human body is a specific condition that lends itself uniquely to an evaluation through simulation modeling. Results of such procedures indicate that efficient thermal averaging over all body tissues of locally deposited RFEM energy will probably occur through convective heat transfer via blood flow."
3. If a given carrier frequency is modulated at a depth of 50 percent or greater at frequencies between 3 and 100 Hz, the NCRP recommendation is to limit the exposure criteria for occupational exposure to those that apply to exposure of the general public. Such a practice may prove to be acceptable in the long run, but one cannot

know with certainty whether this action is justified or whether the general population standard will guarantee the absence of biological effects when exposures occur to modulation frequencies between 3 and 100 Hz.
4. One can still conform to a whole-body averaged SAR of 0.4 W/kg (averaged over 6 or 30 min) while exposing subjects to extremely high peak powers. We need to develop a limit on peak power. This is a very important research project.
5. The use of mobile and hand held transceivers by members of the general public allows exposures that are in excess of the general population standard, but not above the occupational guidelines, as long as the user does not expose others to levels exceeding the general population standard. I have difficulty understanding how the average citizen will comply with the intent of this instruction.
6. With the obvious curtailment of research funds by the federal government, opportunities for further investigation of mechanisms of interaction between RFEM energy and biological systems will undoubtedly be limited. There is a practical need to progress from our present macroscopic understanding of the basis on which exposure limits have been set to a better understanding of the effects of fields internal to the irradiated organism, that is, we need to understand these matters on a microscopic scale. We seem to be headed toward *de minimis* research activity in the biological effects and dosimetry of RF radiation in the midst of explosive activity in the ELF region. This being the case perhaps as a practical matter we should use whatever research monies are available to confirm the validity of the many underlying assumptions contained in the ANSI and NCRP RFPGs.
7. The ANSI-C95.1-1982 Standard (30) and the NCRP Report No. 86 (31) represent a significant improvement over what we had in earlier times. But given the many qualifying remarks contained in the documents, the many conditions that require professional assistance in their interpretation and application, and conditions that cannot be evaluated for the lack of data, I am fearful that most of the qualifications may be forgotten, and predictable inertia may return due to a sense of satisfaction with our present tools.

For my part I wish to thank you for your attention this morning. I will continue to work with all of you on these issues so that sometime in the distant future, perhaps at an NCRP meeting, we may display the standard cosmology in a way that more fully recognizes our joint efforts in radiation protection. If one looks carefully at Figure 1 you will see that I estimate the time for the announcement by NCRP of a grand unified basis of radiation protection (GURP) from both ionizing and nonionizing radiations will occur at approximately 10^{18} seconds after the big bang.

References

1. NCRP. National Council on Radiation Protection and Measurements. *Radiofrequency Magnetic Fields: Properties, Quantities and Units, Biophysical Interaction and Measurements*, NCRP Report No. 67 (National Council on Radiation Protection and Measurements, Bethesda, Maryland) (1981).
2. Turner, Michael S. "Cosmology and particle physics," p. 30 in the *Early Universe*, Unruh, G. H. and Semenoff, G. W., ed. (D. Reidel Publishing Company, P.O. Box 173300 AA Dordrecht, Holland) (1988).
3. Penzias, A. and Wilson. "Determination of the microwave spectrum of galactic radiation," J. Appl. Phys., **146,** 666 (1966).
4. Turner, Michael S. "Cosmology and particle physics," p. 42 in the *Early Universe*, Unruh, G. H. and Semenoff, G. W., ed. (D. Reidel Publishing Company, P.O. Box 173300 AA Dordrecht, Holland) (1988).
5. Hirsch, F.G. "Microwave cataracts," presented at NSAE Conference on Industrial Health, Cincinnati, Ohio, April 25, 1952. Also Hirsch, F.G. and Parker, J.T., "Bilateral lenticular opacities occurring in a technician operating a microwave generator," AMA Arch. Ind. Health, Vol 6, page 512–517, (December, 1952).
6. Schwan, H.P. and Li, K., "The mechanism of absorption of ultrahigh frequency electromagnetic energy in tissues, as related to the problem of tolerance dosage," IRE Trans. of Medical Electronics, Vol. ME-4, page 45–49, (February 1956).
7. Central Safety Committee, Bell Telephone Laboratories, Nov. 1953.
8. Vosborgh, B.L., "Problems which are challenging investigators in industry," IRE Trans. on Medical Electronics, Vol. ME-4, page 5–7, (February, 1956). Also: Vosborgh, B.L. "Recommended tolerance levels of MW energy; current views of the General Electric Company's health and hygiene service," page 118–125 in *Proceedings 2nd Annual Tri-Service Conference on Biological Effects of Microwave Energy*, RADC, Griffiss AFB, New York (July, 8–10, 1958). July 8–10, 1958.
9. "Symposium on physiologic and pathologic effects of microwaves," Mayo Foundation House, Rochester, Minnesota (September 23–24, 1955).
10. Schwan, H.P., "The physiological basis of injury," page 60–63 in *Proceedings 1st Annual Tri-Service Conference on Biological Hazards of Microwave Radiation*, RADC Griffiss AFB, New York, July 15–16, 1957.
11. Frey, A.H., "Auditory system response to modulated electromagnetic energy," J. Appl. Phys. **17,** 689–692.
12. USAF. United States Air Force. "Microwave radiation hazards," Rome, A.F. Depot, Griffiths AFB, New York, Urgent Action Tech. Order, 31-1-511, June 17, 1957.

13. "Biological effects of radio frequency energies," page 94–103 in *Proceedings 1st Annual Tri-Service Conference on Biological hazards of Microwave Radiation*, RADC Griffiss AFB, New York July 15–16, 1957.
14. Temporary Sanitary Rules for Working with Centimeter Waves, Ministry of Health Protection of the USSR, 1958, (1961).
15. The Irradiation of the US Embassy in Moscow was Reportedly Known in the 1950's, but Documented Much Later. Refer to References No. 27 and 28.
16. ANSI. American National Standards Institute. Committee 95.1 Safety Levels of Electromagnetic Radiation With Respect to Personnel, C95.1, New York (1966).
17. Hearings Before the Committee on Commerce United States Senate, Ninetieth Congress, Serial No. 90-49, US Government Printing Office, Washington, D.C., May 1968.
18. Technical Electronic Products Radiation Safety Standards Committee (TEPRSSC) established under the Radiation Control for Health and Safety Set of 1968 (Public Law 90-602).
19. Radiation Control for Health and Safety Act of 1968. (Public Law 90-602). See Annual Reports to the Congress on the Administration of DL 90-602. Bureau of Radiological Health, Washington, D.C.
20. *Proceedings on the Symposium on the Biological Effects and Health Implications of Microwave Radiation*, Cleary, S. F. ed., Richmond, VA, September 1969, U.S. Department of HEW, June, 1970.
21. Hirsch, F. G., "Microwave cataracts—a case report reevaluated," *Electronic Product Radiation and the Health Physicist, Proceedings of the 4th Annual Symposium of the Health Physics Society*, Louisville, Kentucky, Jan. 28–30, 1970 HEW publication BRH/DEP 70-26.
22. Radiation Control for Health and Safety Hearings before the Committee on Commerce, U.S. Senate, Ninety-third Congress, Serial No. 93-24, U.S. Govt Printing Office, Washington, D.C. 1973.
23. Presentations of Consumers Union before the Technical Electronic Products Radiation Safety Standards Committee (TEPRSSC), Washington, D.C. 1973. Presentations of American Home Appliances Manufacturers before the Technical Electronic Products Radiation Safety Standards Committee (TEPRSSC), Washington, D.C. 1973.
24. NAS. National Academy of Sciences. "Biosphere effects of extremely low frequency radiation" (Sanguine/Seafarer), Washington, D.C. 1974–1977.
25. ANSI. American National Standards Institute. Safety Levels of Electromagnetic Radiation with Respect to Personnel, C95.1, New York, NY 1974.
26. USAF. United States Air Force. Radiofrequency Radiation Health Hazard Control, AFR 161-42, Washington, D.C., 1975.
27. U.S. Senate Microwave Irradiation of the U.S. Embassy in Moscow. Hearings before the Committee on Commerce, Science, and Transportation Committee Print 43-949 Government Printing Office, Washington, D.C. 1979.
28. Lilienfield, A.M., Tonascia, J., Tonascia S., Libauer, C.H., Cauthen, G.M., Markowitz, J.A., and Weida, S. *Foreign Service Health Status Study-*

Evaluation of Health Status of Foreign Service and Other Employees From Selected Eastern European Posts, Final Report Contract No. 6025-619073 Department of State, Washington, D.C. 1978.
29. WHO. World Health Organization. *Environmental Health Criteria for Radiowaves in the Frequency Range from 100 kHz to 300 GHZ (Radiofrequency and Microwaves).* WHO Environmental Criteria Program (United Nations, New York) (1981).
30. ANSI. American National Standards Institute. *Safety Levels with Respect to Human Exposure to Radio Frequency Electromagnetic Fields, 300 kHz to 100 GHZ,* Report No. ANSI C95.1-(The Institute of Electrical and Electronics Engineers Inc., New York (1982).
31. NCRP. National Council on Radiation Protection and Measurements. *Biological Effects and Exposure Criteria for Radiofrequency Electromagnetic Fields,* NCRP Report No. 86 (National Council on Radiation Protection and Measurements, Bethesda, Maryland) (1986).

Discussion

DAVE JANES (EPA): Thanks for the walk down memory lane. One of the things that occurs to me that's not generally recognized is that the early standard, Dr. Schwan's, is essentially based on a specific absorption rate working backwards up from the basal metabolic rate to give you an exposure standard. And the other piece, since we're taking this walk down memory lane, we heard yesterday from Paul Slovic about stigma and the concern that electrical engineers have with the word radiation associated with this modality. And there's another side to that coin, non-ionizing connotes with non-problem, and in some sectors, non-funding.

GEORGE WILKENING: I agree. Thank you.

JOHN CAMERON (University of Wisconsin): Is the NCRP committee or any other committee looking into the effects of intense pulsed magnetic fields? Especially on the brain, which would respond quite dramatically to these fields?

GEORGE WILKENING: I think Mahlum's Committee on magnetic fields is considering it.

JOHN CAMERON: Is that an NCRP committee?

GEORGE WILKENING: That's an NCRP committee. It's nearly finished; it's been in the works quite awhile. They are considering pulsed magnetic fields and electromagnetic pulses (EMP).

WESLEY NYBORG (University of Vermont): Just a comment that the problem of localized heating which you mentioned rises with ultrasound too, in that case from focused ultrasound. Certainly there is not nearly enough information available about the biological consequences of heating just a small region of the body.

The State of the Art in Measuring Electromagnetic Fields for Hazard Assessment

Richard A. Tell
Richard Tell Associates, Inc.
Las Vegas, NV

Background

Electromagnetic fields, as they relate to human exposure, are evaluated through analyses or measurements. In many cases, when it becomes impractical to compute the expected fields from particular sources, field measurements are the only expedient method of obtaining the required information. Also, in particularly complicated exposure situations, it may be difficult or impossible to theoretically determine the level of exposure over the period of time that an individual is actually exposed. Thus the challenge of the measurement task is to accurately determine the electromagnetic fields created by a diverse variety of sources. The sources themselves, because of their individual character, can require substantially different measurement approaches to characterizing the exposure which they create.

Some of the sources of electromagnetic fields which are commonly found to be the subject of exposure surveys include:

(1) Broadcasting stations in the AM and FM radio service and television service.
(2) Satellite earth stations used for communications with orbiting satellites in which signals are transmitted into space using highly directive antennas.

(3) Radars which are used for surveillance and tracking operations and which normally use a pulsed signal.
(4) Microwave communication links, more commonly referred to as microwave repeater stations, which are used for relaying telephone conversations and other data from point to point.
(5) Dielectric heat sealing machines which employ the principal of heating of materials through the dielectric losses in materials subjected to very intense RF fields.
(6) Microwave ovens, though not usually the home type, used for large scale, commercial processing of food products and drying operations.
(7) Video display terminals (VDTs) which, because of the electronic circuits used to control the deflection of the electron beam inside the cathode ray tube (CRT) to form images on the screen, produce electric and magnetic fields in their vicinity.
(8) Electric power lines, including higher voltage power transmission lines and lower voltage distribution lines, which generate both electric fields due to the voltage of the line and magnetic fields due to the current flowing through the lines.
(9) RF induction heaters which make use of extremely high intensity magnetic fields to induce eddy currents in materials that produce high temperatures.
(10) Shortwave radio stations, including international broadcast, military communications and amateur radio systems.

In the case of each of these sources, to name only some, electromagnetic fields are created due to the operation of the equipment. In some cases, these fields are relatively wide spread resulting in so-called far field exposures having approximately uniform exposure over the dimensions of the body. In others, the sources are typically smaller physically and tend to cause near-field exposures wherein the fields are very non-uniform over the body and are characterized by rapid changes as a function of distance from the source.

In addition to these differences in field uniformity, the presence of the individual performing the measurements or of the person for whom exposure data is desired, will typically alter, or perturb, the fields present. When evaluating the electric fields found beneath electric power lines and near VDTs it is common practice to specify whether one is describing the perturbed field or the un-perturbed field, *i.e.*, the field without the influence of the individual present.

Electromagnetic fields have a wide range of waveforms depending on their use. For example, a pulsed radar signal typically will exhibit a very high peak to average field strength ratio due to the use of very narrow pulse widths but very high peak powers. In the case of normal AM radio broadcasting, again the average value of the field will be dependent on the programming material and adjustments to the transmitter. The waveform of some fields may be of interest because of their rapid changes in time which, for magnetic fields, may be important from the standpoint of induced currents in the body.

The measurement task is made more complex because of the very wide range of frequencies applicable to the above listed sources, a factor of approximately 10^9 times in frequency. These differences play an important role in determining how the field couples to the human body and the degree to which energy may be extracted from the field by the tissues of the body. The rate of energy absorption is called the specific absorption rate (SAR) and is measured in units of watts per kilogram of body mass (W/kg). The additional complication of field polarization and insuring that the field measurement has properly taken into account all of the relevant energy of the field adds to the long list of potential exposure conditions which confront the individual performing an electromagnetic field survey. Thus, there are a number of issues which must be appreciated to avoid pitfalls in the measurement process and the selection of appropriate instrumentation for the job. Fortunately, present-day instrumentation lends itself quite well to many of the measurement tasks confronting the health physicist, industrial hygienist or others whose task is to determine electromagnetic fields in the course of hazard assessments.

Physical Parameters Often Used in Electromagnetic Field Evaluations

The magnitudes of the electric (E) and magnetic (H) field components, and sometimes their squares, are the primary field parameters which enter into any electromagnetic field hazards assessment. The electric and magnetic fields are related to one another at long distances from a source, in the so-called far-field region, through what is called the intrinsic impedance of free space (air) as given in equation [1]:

$$E(V/m)/H(A/m) = 377 \text{ ohms} \qquad [1]$$

The field magnitudes are expressed in terms of volts per meter (V/m) and amperes per meter (A/m) for electric and magnetic fields respectively. Both of these fields are commonly expressed in terms of an equivalent power density of a planar electromagnetic wave, *i.e.*, in the equivalent of free space in the far field of a source. In this case, the plane wave equivalent power density (S) is given by equation [2]:

$$S(W/m^2) = [E(V/m)]^2/377 = 377*[H(A/m)]^2 \qquad [2]$$

Power density is more commonly expressed in units of milliwatts per square centimeter (mW/cm^2). Instruments designed for electromagnetic field hazard surveys are typically calibrated in terms of either field strength, field strength squared units or plane wave equivalent power density. As will be seen, many instruments use detectors which are actually responsive to the square of the field strength, either electric or magnetic, but provide meter readings in plane wave equivalent power density; the meter indications are obtained via an electronic manipulation of equation [2].

Other parameters of importance are the frequency or frequencies of the principal source(s) and the duration of the exposure. Virtually all present day standards for electromagnetic field exposures are frequency dependent; this dependence takes into account the frequency response of the human body in terms of energy extraction from the incident fields. Because the body acts much like a radio antenna in that it absorbs energy from the field better at certain frequencies, for example the body resonance range, most recommended limits for exposure levels vary in accordance with this frequency selectivity of the body with the most stringent controls in the body resonance range and less restrictive levels at frequencies both lower and higher than the body resonance range. The implication of this is that when performing a survey of exposure, the frequency of the field is necessary to relate to permitted exposure levels and the instrument being used for the measurements must be capable of accurate response at the frequency of exposure.

Because most recommended exposure limits for electromagnetic fields are based on some specified averaging time, instruments which contain sophisticated microprocessor circuitry, can provide insight to field values averaged over various times. These types of instruments find application in situations having highly time varying exposures such as radio tower climbing.

Broadband and Narrowband Measurement Approaches

Broadband instruments

In terms of convenience, broadband instruments are considerably more popular than narrowband instruments for evaluating exposure to electromagnetic fields. Broadband instruments, by their name, are devices which are responsive over a wide range of frequencies, possessing near flat responses (independent of frequency). Thus, as long as the source of interest operates within the frequency passband of the instrument, the indicated field magnitudes will be measures of the exposure. The fact that the instrument is broadband does not, however, remove the responsibility from the user to ascertain that the instrument bandpass includes the pertinent frequency. This, in general, means that the user should determine, either through contact with operators of the source or via an independent measurement, the frequency of the source.

Broadband instruments take two forms; those that are isotropic, *i.e.*, their output does not change materially with orientation of the sensing probe in the exposure field, and those that are sensitive to one polarization of the field, *i.e.*, they require that the sensing element of the probe, or instrument, be aligned with the incident field polarization or that orthogonal measurements be taken of the field strength such that the resultant field magnitude can be determined.

Isotropic probes consist of three mutually orthogonal detecting elements, connected electronically in such a manner that the output of the instrument to which the probe is connected becomes essentially independent of orientation in the field. This property permits the user to ignore the absolute polarization of the fields but rather concentrate on the collection of exposure data at different points. In reality, isotropic probes are never perfectly isotropic but exhibit some degree of ellipticity or slight variation (commonly this variation is no greater than a few tenths of a decibel) in indicated field strength when rotated in a linearly polarized electromagnetic field. Typically, each sensitive probe element detects the component of the applicable field parameter, be it the electric or magnetic field, and the attached instrument electronically sums the outputs of the three probe elements to produce an output proportional to the resultant vector magnitude of the field parameter. Normally such summing is done by processing the output of either diode or thermocouple detectors incorporated in the sensing probe elements. Thus, no information on the electrical phase of the fields is retained. Recent work at the National Bureau of Standards (NBS) to develop a so-called Poynting vector meter in which both the electric and magnetic fields as well as the phase angle between them has been described by Kanda (1). Aslan (2) has described the development of an isotropic probe using linear arrays of thin-film thermocouples as the detection elements. Thermocouples exhibit the useful property of being average responding detectors; when placed in a modulated field, such as a pulsed radar field, a thermocouple detector will accurately respond to the average value of the field squared or average plane-wave equivalent power density. This same property, *i.e.*, true root-mean-square (RMS) detection, of thermocouples, makes them useful in multiple frequency electromagnetic field environments since they can accurately respond to the sum of the average fields present at the probe.

Isotropic probes using diodes as detectors have been developed by the National Bureau of Standards (3) and designs derived from these initial NBS prototype instruments have now been commercialized. Diode detectors tend to be peak detectors rather than average detectors unless operated within limited field intensity ranges where they can exhibit so-called square law responses, *i.e.*, they will respond correctly to the power density or square of the fields. In some instruments, special circuits are used to correct for the non-square law operation of diode detectors; this technique works well for single frequencies but cannot properly correct readings when multiple signals are being measured. While diode detectors have certain disadvantages, they also have several advantages over thermocouples in terms of durability, ability to withstand overloads from strong fields, better thermal stability and high sensitivity.

The most common type of isotropic survey instruments consist of separate probe assemblies which are connected to an associated electronics readout package via a cable. Other forms of survey instruments have single or multiple sensing probes mounted directly on the instrument package. Those instruments equipped with one probe element are responsive to only one field polarization component at a time. The absence of a connecting cable between the probe and the readout package, in some cases, is responsible for superior performance when used in low frequency fields; it is common for cables to "pick up" or couple with the electric field at lower frequencies, typically below a few megahertz, often times leading to erroneous indications of field strength. Cableless instruments, or ones equipped with non-conductive, fiber optic cables, are preferred for electric field measurements at low frequencies. Mantiply (4) has recently provided useful information on the performance characteristics of broadband instruments when employed in different types of electromagnetic environments.

Present-day broadband instruments exhibit flat frequency responses over large frequency ranges depending on their specific application. For example, electric field probes can be purchased with manufacturer specified passbands as wide as 200 kHz to 40 GHz. Magnetic field probes typically offer narrower bandpass characteristics such as 0.5 MHz to 300 MHz. The reduced bandpass of magnetic field probes is related to the use of small loop sensors which tend to exhibit high frequency resonances leading to erroneously high responses at some specific frequencies. Achieving such wide bandpasses for electric field probes is accomplished by special loading of the probe elements such as to maximize their flatness of response.

Narrowband instruments

Broadband instruments, while convenient for conducting area surveys and rapidly assessing exposure levels of electromagnetic fields, do not yield information on the frequency of the fields illuminating the measurement point. In many situations this may not be necessary but because of the frequency dependence of most RF field protection guides, in areas of mixed frequency fields where more than one source is present, narrowband equipment may be necessary to correctly evaluate the exposure. For example, if two frequencies are present for which different exposure limits apply, it is necessary to individually determine the contribution that each field makes to the total exposure to declare whether the protection guide is being exceeded. This, in turn, requires that the measuring equipment have selectivity for distinguishing the various frequencies that are present. While narrowband instruments are inherently more complex to apply in field measurements than their simpler broadband counterparts, they serve an essential role in many situations where frequency resolution and measurement sensitivity are strong considerations.

Conventionally, narrowband field measurements are accomplished by using a tuneable receiver connected to a calibrated receiving antenna. By measuring the RF voltage output of the antenna with the receiver, and knowing the antenna's calibration factor in terms of output voltage as a function of field strength, the incident field strength can be determined. Dipole antennas are commonly used as the receiving antenna for frequencies below 1 GHz for electric fields and loop antennas for measurement of magnetic fields less than 30 MHz. An alternative to the tuneable receiver, in which each emission frequency is individually tuned in, is the spectrum analyzer. The spectrum analyzer repetitively scans between a lower and upper frequency producing a graphical display of the measured, received signal powers, or voltages, delivered from the reception antenna. Spectrum analyzers permit rapid assessment of the relative signal amplitudes across a wide range of frequencies and are particularly effective when the frequency of a specific source is unknown or the presence of the field is intermittent.

In either case, both tuneable receivers and spectrum analyzers, along with their companion sensing antennas, are not as portable as broadband types of instruments. In many cases, where it is important to ascertain the spatial distribution of the fields, it is very inconvenient or simply impractical to conduct the measurement using the typically large antennas employed with narrowband instruments. This is a serious consideration when performing field measurements under difficult circumstances such as on tall broadcast antenna towers and in other situations where it is impractical to carry narrowband equipment but, nevertheless, where the exposure is produced by a multiplicity of signals across the spectrum. In these cases, the measurement procedure must take into account this fact and arrangements must commonly be made to operate the various sources individually so that their individual contributions can be assessed with a broadband type of instrument. In some cases, the arrangement of turning off some sources even for short periods of time may prove impractical, such as at large multiple broadcast transmitting facilities. In this case, one alternative is the use of a relatively recently introduced measurement probe which incorporates a frequency response which accounts for the frequency variation of the RF field exposure criteria (5). This type of instrument weights the intensity of the fields according to frequency such that the instrument readout is indicated in terms of the percentage of the exposure criteria permitted; for example, if the instrument indicates 100 percent, it means that the frequency weighted exposure field is equal to the permitted value. As with any broadband type of instrument, however, it does not indicate how the indicated resultant field magnitude is distributed across the frequency spectrum.

Assemblies of narrowband instruments may be configured into sophisticated, automated measurement systems which rely upon computer control for collecting the intended field strength data. Several of these systems have been configured for applications in environmental monitoring of ambient RF fields (6, 7). In these systems, spectrum analyzers interfaced to small computers have been used with various types of antennas for measurements across wide frequency ranges. Through the use of electronic switching techniques for connecting different antennas, the entire process can be essentially automated. In addition to the repeatability of the measurement process offered by such systems, specialized measurements may be accomplished such as retention of instantaneous peak field strengths such as might occur with intermittent sources or pulsed radars, automatic correction of the field strength data for system characteristics like antenna calibration factors which vary as a function of frequency and signal averaging of the typically time varying field strength data. An example of the power of computer assisted systems is the ability to measure the time varying signal levels associated with rotating radar antennas and the real-time calculation of average field strength over a complete rotation of the radar's antenna. These sophisticated measurement systems are generally large in size, often being installed in vehicles for portability from one location to another. Thus, while they represent extremely powerful tools for evaluating electromagnetic field exposure environments, they can present significant problems of implementation in many high level RF field situations.

Special Adaptations or Applications of Instruments

A particularly noteworthy development in the last several years has been the introduction of data processing hardware which can be configured with portable RF survey instruments to simplify the data collection task. These devices, which have taken the form of microprocessor based data-loggers or analog devices for simply integrating exposure over some prescribed time interval, now allow a direct assessment of the time-averaged exposure to electromagnetic fields. The power of such devices, which normally operate by processing the recorder output signal from the associated field strength measuring instrument, is that one can use them to actively manage personnel exposures in areas where the exposure levels may reach substantial values momentarily, yet by constantly monitoring the time-averaged value, the exposed individual may remove themselves from the field to keep the average level to within acceptable limits. Tell (8) has described the use of data-logging techniques during antenna tower climbing work. An advantage to data-logging devices with internal memories is that the data can be down-loaded to a computer subsequent to the measurement, examined and analyzed. These devices allow for the measurement of spatially averaged values of electromagnetic fields as well since the probe can be moved through the space being surveyed at a uniform speed so that the resulting averaged value is representative of the field's spatial average. This type of measurement is important in attempting to assess the specific absorption rate of the body when immersed in non-uniform (partial body exposure) fields.

The measurement of 60 Hz electric fields associated with power lines, and more recently with the development of instrumentation for the measurement of VDT electromagnetic field emissions, has made use of so-called displacement current sensors. These devices rely on the fact that an electric field impressed on a pair of closely spaced, parallel, conductive plates which are electrically connected together, will cause a small current, the displacement current, to flow between the plates. This displacement current flow represents the rearrangement of electric charge on the two plates such that the electric field between the shorted plates remains at zero. By calibrating the displacement current as a function of externally applied electric field strength, the device can be used to measure field strength. Such meter designs are sometimes referred to as free-body meters in that they lend themselves to the measurement of unperturbed electric fields in space by not being referenced, or electrically connected, to ground. The usual absence of electrically conductive cables provides for minimum field pickup effect and, when the meter is supported in the field via a non-conductive means, will yield a measure of the unperturbed field. Commonly, these free-body meters are constructed as integral units with the associated electronics being built inside the sensor assembly.

The problem of electric field pickup by conducting cables, mentioned above, can also be a problem with low frequency, narrowband measurements in which electric field antennas are used. A relatively recent use of fiber optic techniques for conveying the signal from an electric field responding antenna has been described by the Environmental Protection Agency (9). A spherical dipolar antenna, consisting of two hemispherical shells separated by a thin insulator contain the necessary electronics to convert the RF signals induced on the two halves of the dipole into an amplitude modulated light beam which is transmitted through a plastic, fiber optic cable to a remote electronics package. The electronics package is used to demodulate the light beam and reconstruct the original RF signal spectrum induced on the antenna. The fiber optic cables effectively electrically isolate the antenna from its environment, thereby preventing the troublesome artifacts of metallic cable field pickup. Optical isolation methods for narrowband antennas are still in the experimental stage and expensive but the use of pulsed light beams to transmit the equivalent of the recorder output voltage from portable electric field sensing instruments has proved technically feasible and is commercially available for specialized applications.

An area of instrumentation highly developed in the ionizing world of radiation, personal dosimetry, has not been successfully developed for the general case of electromagnetic fields. This has been the case because of the great difficulty in attempting to relate radiofrequency field strengths determined very close to the surface of the body to equivalent, unperturbed values from which SAR may be derived. At lower frequencies, however, personal dosimeters have been developed for use in power frequency fields, *i.e.*, 50 and 60 Hz. At these frequencies, especially for magnetic fields, body perturbation effects can be accounted for in a simpler manner than for higher radio frequencies. Some of these dosimeters contain advanced digital circuitry which allows the retention of data on exposure levels versus time of day (10). Hence, very complete exposure histories can be determined for subjects wearing these devices.

References

1. Kanda, M. "An electromagnetic near-field sensor for simultaneous electric and magnetic-field measurements," page 102–110 in *IEEE Transactions on Electromagnetic Compatibility*, Vol. EMC-26 (Institute of Electrical and Electron Engineers, New York) (1984).
2. Aslan, E. "Broadband isotropic electromagnetic radiation monitor," page 411–424 in *IEEE Transactions on Instrumentation and Measurement*, Vol. IM-21, No. 4 (Institute of Electrical and Electron Engineers, New York) (1972).
3. Bowman, R.R. *Quantifying Hazardous Electromagnetic Fields: Practical Considerations.* National Bureau of Standards Technical Note 389 (National Technical Information Service, Springfield, Virginia) (1970).
4. Mantiply, E.D. "Characteristics of broadband radiofrequency field strength meters," in *Proceedings of IEEE Engineering in Medicine and Biology Society 10th Annual International Conference*, New Orleans, LA, November 4–7 (Institute of Electrical and Electron Engineers, New York) (1988).
5. Aslan, E. "An ANSI radiation protection guide conformal probe," Microwave Journal, April (1983).
6. Tell, R.A., Hankin, N.N., Nelson, J.C., Athey, T.W. and Janes, D.E. "An automated measurement system for determining environmental radiofrequency field intensities: II," page 203–313 in *Measurements for the Safe Use of Radiation*, Fivosinsky, S.P., Ed., National Bureau of Standards Special Publication 456, (National Technical Information Service, Springfield, Virginia) (1976).
7. Tell, R.A. "Instrumentation for measurement of electromagnetic fields: equipment, calibration and selected applications," in *Biological Effects and Dosimetry of Nonionizing Radiation.* Grandolfo, M., Michaelson, S. and Rindi, A. Eds., NATO Advance Study Institute Series, Series A: *Life Sciences*, Vol. 49, (Plenum Publishing Company, New York) (1983).
8. Tell, R.A. "Real-time data averaging for determining human RF exposure," page 388–394 in *Proceedings 40th Annual Broadcast Engineering Conference*, National Association of Broadcasters, Dallas, Texas, April 12–16, (1986).
9. EPA. Environmental Protection Agency. "An investigation of radiofrequency radiation levels on Healy Heights, Portland, Oregon, July 28–August 1, 1986," Technical Report prepared by Electromagnetics Branch, Office of Radiation Programs, (U.S. EPA, Las Vegas, Nevada) (1987).
10. EPRI. Electric Power Research Institute. EMDEX (electric and magnetic field digital exposure) system, Technical Brief RP799-16 (Electric Power Research Institute, Environment Division, Palo Alto, California) (1987).

Protection in Medical Ultrasound—Modest Progress in a 'Low-Risk' Field

Paul L. Carson
J. Brian Fowlkes
University of Michigan Medical Center
Ann Arbor, MI

Abstract

Recent efforts in the U.S. have led to measurement, assessment and modest reporting of ultrasound exposure levels from medical ultrasound systems. A great impediment, as well as aid, to protection has been the popularized desire and belief that diagnostic ultrasound is, and must be, totally without risk to the patient. This risk perception and decision has helped maintain large scale use of ultrasound and has, indeed, helped keep acoustic emissions at relatively low levels. However, the belief in the safety of ultrasound has impeded the measurement and reporting of exposure parameters and threatened research and development in bioeffects and dosimetry and development of newer, possibly invasive techniques.

Many discussions have occurred over the past few years regarding the need and mechanism for FDA approval of Doppler ultrasound for obstetrical applications. Recently, discussion has centered on real time labeling of spatial peak, temporal average intensity, I_{SPTA}, and possibly one temporal peak field quantity. This might be required only when these quantities exceed certain levels. Real time labeling means on-screen labeling, or possibly control knob labeling in simpler systems. I_{SPTA} has been discussed as the single labeling parameter because of the expectation that, among possible bioeffects mechanisms and their currently known thresholds, we are closest to the thermal thresholds. The possibility of cavitation is the basis for some wishing to label a temporal peak quantity such as the maximum rarefactional (negative) pressure p_r. The fact that I_{SPTA} is only a very indirect indicator of expected temperature increases has been a disadvantage. Temperature rise depends strongly on beam width and frequency. It is now being discussed by NCRP Committee 66 to recommend real time labeling of a quantity like the thermal index, given by:

$TI = W_o/W_{deg}$,

where W_o = Acoustic power (mW) and W_{deg} = estimated power to raise tissue temperature 1° C. The state-of-the-art in ultrasound protection is summarized by the observation that the field is changing rapidly; many will not recognize most of the figures in this report.

Introduction

Maintenance of the relative safety of ultrasound and of the perception of safety for frequent use in sensitive situations such as pregnancy, is more important for medical care and is a major responsibility for protection efforts. However, the desire for a guarantee of absolute safety has tended to impede the growth in our knowledge of ultrasound bioeffects and ultrasound exposures. It has been necessary to regularly defend the need for further bioeffects and dosimetry research. It has even been argued by some that there is little need for exposure measurements, and it has been difficult to obtain and publicize output information from many companies. It is the position of many in industry, explicitly accepted by the FDA, that details of transmitted waveform *shapes* are proprietary information.

The desire to consider ultrasound totally safe for essentially any use has also impeded the development of those advanced medical ultrasound diagnostic capabilities that might bring some measure of risk, or at least bring us closer to damage thresholds.

Despite these difficulties caused by the assumed or desired absolute safety, considerable progress is being made on protection in medical ultrasound and the field is an extremely rich and diverse one. It is being stimulated largely by the expanding use and technology of medical ultrasound. Marketing planners estimate that in the early 1990's the volume of ultrasound equipment sales worldwide will exceed that of

TABLE 1—*Estimated in-water and derated or "In Situ" Spatial Peak-Temporal Average Intensity*

Use	I_{SPTA} (mW/cm^2)	
	Derated	Water
Cardiac	430	730
Peripheral Vessel	720	1500
Ophthalmic	17	68
Fetal Imaging and "Other"	94	180

any other imaging modality such as CT, MRI, nuclear imaging or x ray systems.

Labeling and Emission Guidelines

I will not discuss international standards and other international protection efforts other than to mention that the U.S. maintains its usual oceans-length distance from the main international ultrasound standards efforts in the International Electrotechnical Commission (IEC). That is, a small number of U.S. representatives attended the IEC meetings somewhat sporadically and there is little U.S. effort to prepare for IEC meetings. That is despite the excellent organizational efforts by the U.S. technical advisor for IEC Technical Committee 87, Dr. Peter Edmonds. I would suggest that at the time of consolidation of the European Economic Community into a single economic entity, as scheduled for 1992, it would be appropriate to reorganize IEC representation in terms of a more balanced representation of the major economic blocks. Meetings also should be conducted in North America more often.

Ultrasound regulation in the U.S. has continued to be conducted under the medical device act of 1976, in which approval of most equipment has been obtained through the 510k process. Approval for marketing under that process requires demonstration that the equipment is equivalent in safety and effectiveness to pre-1976 ultrasound devices. Equivalence has been defined as having intensities I_{SPTA}, I_{SPPA} and I_m less than certain values. The spatial peak, temporal average values are listed in Table 1. They are given in terms of measured values in water and the derated or so-called *in situ* values, as calculated by a homogeneous attenuation model with attenuation coefficient 0.3 dB cm^{-1} MHz^{-1}. With some flexible interpretation by the FDA this approach has essentially worked, with some occasional long delays. A few Doppler ultrasound techniques are only just now beginning to be approved for fetal use.

1. "Power up" and "new exam" default to appropriate typical level.
2. Reminders and required override action when output is set above the default setting.
3. Display of exposure, I_{spta} and P_r, or some other appropriate temporal average and peak quantity.

FOOTNOTES
No need to indicate p_r if it is below some level, 0.5 MPa. Indicate every doubling thereafter.
Perhaps there should be an overall cap on output?

Fig. 1. Labeling Desires Expressed by One FDA Staff Member

Systems are now being reviewed on a case-by-case basis, without standards.

There has been a strong evolution of thought in informal exchanges between FDA representatives, user/scientists represented by the AIUM (and to some extent others such as other medical specialties and the AAPM), and manufacturers, individually and through NEMA. **It was mutually proposed in the summer of 1988 that applications to fetal Doppler, with certain appropriate demonstrations of efficacy will be approved for use if appropriate real time indications of system output can be agreed upon, at least for the higher output levels. It was also agreed that if real time labeling is provided, the current output limit would be removed.** The FDA position as stated informally by one FDA participant in the discussions is summarized in Figure 1. In a manufacturer's response, an output display with only three levels of intensity was proposed as shown in Table 2.

Points of concern in the user/scientist community have been the specification of defacto limits on maximum intensities without an efficacy and safety basis for those intensities. Also, there has been a concern that the so-called *in situ* calculations of intensity are not conservative in important applications such as obstetrical ultrasound, where a major

TABLE 2—*Manufacturer proposal to meet FDA—proposed on-screen or other real time labeling for fetal Doppler devices, with a precision of three levels, in terms of "derated" SPTA intensity:*

Low:	$0 < I_{SPTA,H} < 94$ mW/cm^2	"Fetal"
Medium:	$94 < I_{SPTA,H} < 430$ mW/cm^2	"Cardiac"
High:	$430 < I_{SPTA,H} < 720$ mW/cm^2	"Peripheral Vascular"

Proposed Maximum values of other intensities are the same for each level at $I_{sppa} < 190$ W/cm^2 and $I_m < 310$ W/cm^2 (13 MPa, p_c).

part of the ultrasound propagation path can be relatively nonattenuating fluid. Some alternatives will be discussed, particularly in a following section on protection from thermal effects.

Protection in Relation to Cavitation Mechanisms

There is an intimate relation between the type of effect or mechanism and the protection to be provided. Progress is being made in defining appropriate measurement quantities relevant to both thermal and cavitational mechanisms of damage, the two most significant mechanisms in light of current knowledge. Cavitation is the interaction of ultrasound with gas bubbles. In the presence of bubbles of resonant size, ultrasound might cause significant effects. This is because of the relatively large amplitude of bubble oscillations, due to the extremely large difference in compressibility between the gas and liquid.

At relatively low pressure amplitudes, microscopic bubbles can be increased in size by rectified diffusion. This is caused by both an inflow of gas in the low pressure phase, when the bubble surface is larger, which exceeds the outward diffusion during the bubble compression and a "shell" effect due to a gradient in the dissolved gas concentration near the bubble wall. Thresholds for creating cavitation, when absolutely no cavitation nuclei are present, are extremely high. Possible cavitation nuclei include small stabilized bubbles, gas trapped in crevices and perhaps point discontinuities. Cavitation nuclei are extremely difficult to eliminate in experimental situations, but are extremely difficult to isolate in mammals.

When bubbles are subjected to high amplitude ultrasound, they can collapse in what is termed transient cavitation. In these situations, extremely high temperatures are created in a very small volume, forming free radicals and other chemical reactions. An example is shown in Figure 2, in which chemical reactions during bubble collapse cause light emission from the oxidation of the chemical, luminol.

When pulse echo diagnostic ultrasound as well as lithotripsy waves propagate through water, the nonlinear propagation and minimal attenuation of the generated harmonics produce short rise times and large peak positive pressure amplitudes. A typical spark gap lithotripsy pressure waveform is shown in Figure 3.

Some early fears have been reduced regarding the cavitation potential in diagnostic systems of high values of peak positive pressure and rapid rise times generated by nonlinear propagation of high pressure pulses. Ayme' et al. (1) have observed that the amplitude of the envelope of the

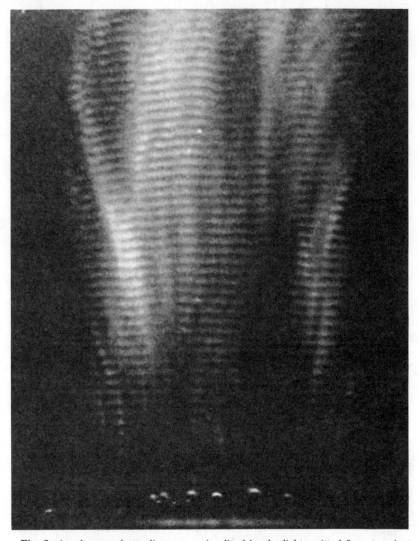

Fig. 2. An ultrasound standing wave, visualized by the light emitted from transient cavitation in water with luminol.

pulse, may be the best indicator of cavitation potential. Here, we interpret that observation as being consistent with the fundamental or low frequency component of the pulse being a good indicator of cavitation potential and with the known decrease in threshold for bubble formation or growth, neglecting resonant effects or maintaining the assumed bub-

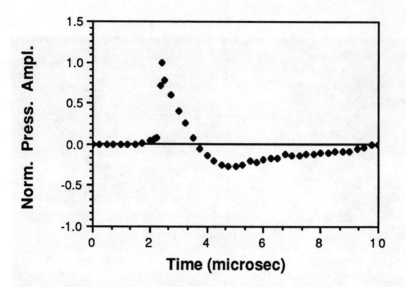

Fig. 3. Pressure waveform at the focus of a spark gap lithotripter.

ble diameter to keep the resonant frequency at a set fraction of the ultrasound frequency. The fairly large frequency dependence of the thresholds for transient cavitation in water containing 1.0 micron diameter bubbles is demonstrated from the results of calculations plotted in Figure 4. For most diagnostic pulses, a quantity which is roughly equivalent to the amplitude of the envelope or harmonic component, and

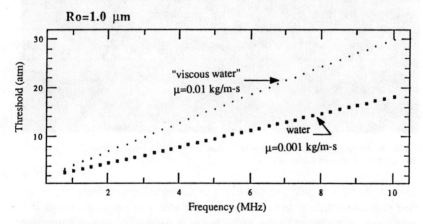

Fig. 4. Frequency dependence of the ultrasound pressure threshold for transient cavitation in water and in a water-like fluid with ten times the viscosity of water (Adapted from Holland and Apfel).[3]

which is easily measurable is the negative pressure amplitude of the pulse(p_r). The dependence of cavitation threshold on the ultrasound frequency is largely ignored by most of the bioeffects statements. For example, comparisons of diagnostic devices with lithotripsy devices (2) talk about the amplitude differences, but ignore the order of magnitude difference in the dominant frequency (typically 120 kHz) of the negative cycle in spark gap lithotripsy pulses.

Typical cavitation bioeffects summaries also fail to adequately emphasize the apparent difficulty of producing cavitation in the mammalian systems at diagnostic frequencies and without unusual conditions of supersaturation. (4, 5) These omissions are based on a sensitivity to the relatively small amount of effort devoted thus far to observing cavitation or its effects *in vivo*. Given the progress in describing the most relevant pulse waveform characteristics for cavitation, it is likely that there will soon be an agreement on what properties should be measured and possibly labeled. The current draft of the AIUM/NEMA acoustic output measurement and labeling standard for diagnostic ultrasound equipment (approved by AIUM) requires labeling of I_{SPPA} and p_r. A recent draft of IEC Working Group 12, TC 87, on acoustic output labeling requirements for diagnostic ultrasonic equipment calls for p_c and p_r, or p_+ and p_- in their symbols. In both documents, these quantities are measured at the point of maximum I_{SPPA}, or the related pulse pressure squared integral in IEC terminology. **Because of the dependence of cavitation threshold on pulse duration and possible amplitude variations within a pulse, the authors believe a pulse average quantity, such as I_{SPPA}, coupled with a printed record of pulse duration, will be a better indicator of cavitation potential than p_r, the best single-cycle indicator.** This is particularly true when comparing the long pulses at long window settings in pulsed Doppler units with the common shorter pulses. It appears that at typical diagnostic frequencies we are further away from known thresholds for frequent cavitation *in vivo* than from thermal effect thresholds.

Potential Temperature Rises in Medical Ultrasound Beams

Scientific bioeffects statements often say "We think ultrasound is relatively safe, go ahead and use it, but, by the way, we don't really know." The protection guidelines and statements should be more operationally useful, such as: "Ultrasound equipment and procedures complying with the protection guidelines provide excellent safety for medically indicated procedures. Output above the guidelines requires a risk/

benefit decision by the physician." An example of an unfortunate state of protection knowledge just 18 months ago was exemplified in the Oct. 87 AIUM thermal bioeffects conclusions(2), which said that for a defined range of typical diagnostic conditions and below 200 mW/cm (2) I_{SPTA}, the temperature will not significantly exceed 1° C. *"However, if the same beam impinges on fetal bone, the local temperature rise may be much higher."* The ambiguity of this potentially important warning is now being alleviated as discussed later.

For obstetrical conditions, the basis of an alternative to the homogeneous attenuation model was presented in NCRP Report No. 74 (6). That model assumed that the ultrasound traverses a fixed thickness of soft tissues overlying the fetus, plus whatever additional amniotic fluid path is required to place the fetal surface at the focal point. Estimated minimum tissue thicknesses overlying the embryo/fetus in the first and third trimesters were presented in Table 2.4 of NCRP report 74 (6), along with conservatively-estimated attenuation coefficients of each tissue. Those estimates resulted in a quite small 1.6 dB minimum attenuation at 3.5MHz during the third trimester and 3.9 dB minimum attentuation during the first trimester. More recently, measurements for 22 patients between 15 and 20 weeks gestation (second trimester) were made of minimum tissue layer thickness potentially overlying the fetus (7, 8). The minimum calculated attenuation of 2.8 dB at 3.5 MHz was proposed as a reasonable minimum attenuation to assume for diagnostic examinations in cases where one does not want to use a more sophisticated relation incorporating the maternal weight.

The differences can be quite large between the homogeneous attentuation model and the fixed tissue path obstetrical model. The two models can be summarized as follows:

The CDRH intensity derating factor is

$$\phi = \exp - (0.23 \cdot 0.3 \cdot f \cdot z)$$
$$= \exp - (0.069 \, f \, z),$$

i.e., the homogeneous tissue attentuation model $H_{.3}$, with attenuation coefficient 0.3 dB cm^{-1} MHz^{-1} and with tissue path distance z. For the proposed second trimester minimum fixed body wall plus amniotic fluid model, the attenuation is:

$$\phi = \exp - (0.23 \cdot 0.8 \cdot f)$$
$$= \exp - (0.18 \, f).$$

In a relatively extreme case of a high frequency, 5MHz, transducer with 7 cm focal length, the attenuation $\phi = 0.089$ and 0.41, respectively, in the homogeneous attenuation model $H_{.3}$ and the 2nd trimester fixed

> ### The Thermal Index
>
> $$TI = W_o/W_{deg}$$
>
> W_o = Acoustic power (mW)
> W_{deg} = Estimated power to raise tissue temperature 1°C

Fig. 5. A Possible Labeling Quantity

attenuation mode, F_2. So the so-called *in situ* intensity by model H_{-3} may underestimate the exposure to the fetus by a factor of 4.5 and is quite inappropriate as an indicator of potential for risk in obstetrical cases.

Calculations of Potential Temperature Rise and Labeling Thereof

Recent calculations and experiments by NCRP Committee 66 (9) and others have included the output powers and SPTA intensities for various transducer diameters, frequencies and focal diameters to produce a 1° C temperature rise in tissues; (10, 6, 11, 12, 13, 14, 2) Most of these works included perfused tissues and several included heating in bone. One of the objections by manufacturers and others of using I_{SPTA} for on-screen labeling has been that it varies dramatically in its potential for increasing temperature, depending on transducer frequency and focal properties. At the 1988 World Federation of Ultrasound in Medicine and Biology meeting and associated AIUM committee meetings, M. Curley and C. Hottinger proposed a calculated *in situ* "absorbed power", which, not surprisingly, tracked very well with expected temperature rise with less dependence on transducer parameters. This quantity utilized tables of calculated data published by the AIUM (2); the latter, in turn, were based on theory developed for the NCRP by its Committee 66 (9). This same committee has proceeded further to generate specific algorithms for deriving a "thermal index", defined in Figure 5.

The thermal index is particularly desirable for on-screen labeling in that it is easily interpreted by physicians responsible for individual examinations. This index also makes real time calculation quite simple for manufacturers, in that they only need to measure the power output at several pulser settings and know the transducer aperture or beam diameter, depending on the exam conditions as described below. A simplified calculation for W_{deg}, a power which is close to, but not exceed-

Adults with normal periostial nerves:
$$W_{deg} \approx 75 \sqrt{D/f}$$

For 1st trimester embryo/fetus:
$$W_{deg} \approx 75\sqrt{d/f}\ 10^{A/10}$$

Where atten. coef. A = 1.0 f (dB),
d(mm) = 6 dB focal diam
D(mm) = transducer diam

Fig. 6. Simplified calculations of the power to produce a maximum temperature rise of 1°C for use in the thermal index, excluding second and third trimester fetal studies (9).

ing, the power expected to produce a maximum temperature rise of 1°C is expected to be proposed (9). The power can nearly always be scaled as the square of the applied voltage.

Figures 6–7 were developed by NCRP Committee SC66, based on its rather extensive tables of calculated thermal effects in soft tissue and more recent calculations in bone (9, 11, 12). W_{deg} represents a power which is close to, but not exceeding the power expected to produce a maximum temperature rise of 1°C.

The results of the thermal calculations in bone are summarized in Figure 8, in which significant, 5°C temperature rises are predicted within 10–20 seconds for a 45 mW beam with 3.5 mm beam diameter (470 mW/cm^2). Equivalent temperature rises were obtained experimentally on mouse skulls with 1.5 W/cm^2 axial intensity, 3.6 MHz and 2.8 mm 6 dB beam diameter. More recent calculations for the adult mouse than those in Figure 8 have included a higher, more realistic, conductivity for bone and finite bone thickness. These calculations show an almost exact correspondence with the experimental data, (15) in terms of temperature rise time and equilibrium temperature for the same acoustic beam parameters.

There are many aspects of this bone temperature rise modeling and experimentation which are yet to be studied and fully understood. For

For ossified, > 13 week, fetuses

$$W_{deg} = 4d\ 10^{A/10},$$

where d(mm) = 6 dB focal diam
and attenuation/MHz is
A(dB)/f(MHz) = 0.5 (3rd trim)
or .75 (2nd trim)

Fig. 7. Simplified calculation of the power to produce a maximum temperature rise 1°C for use in the thermal index in second and third trimester fetal studies (9).

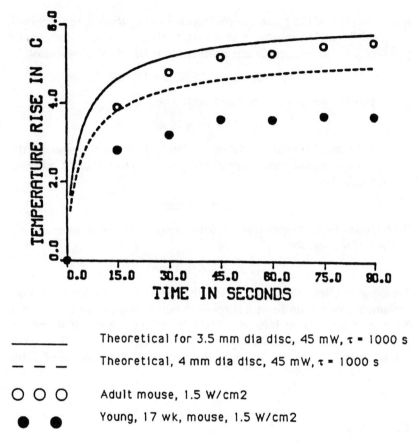

Fig. 8. Calculated and experimental heating of mouse skulls in focused, 3.6 MHz beams (adapted with permission (15).)

example, the lower temperature rise in the young mouse bone is probably due to the smaller amount of ossification, but it has not been quantified nor the relations to human fetal bone clearly drawn. However, the calculations are probably accurate enough that now, for the first time, a calculated temperature rise can give a physician a meaningful estimate of the output relative to that which could cause a significant bioeffect.

The simple guideline of Figure 6 for W_{deg} in bone offers the same simplicity of calculation as the guidelines for soft tissue. While the soft tissue guidelines are not restrictive in relation to current diagnostic equipment, the guidelines for fetal bone would indicate that the temperature rise could exceed 1°C with some of the higher powers and intensities of several of the pulse Doppler ultrasound systems. As an example, a typical 3.5 MHz transducer of 3 mm, 6 dB focal beam width would produce a beam with minimum attenuation in the third trimester of:

$$A = 0.5 \text{ dB/MHz} \times 3.5 \text{ MHz} = 1.75 \text{ dB}$$

the 1° power calculated from Fig. 7 would be:

$$W_{deg} = 12 \times 1.5 = 18 \text{ mW}.$$

For a cylindrical intensity profile, or dividing the total power by the beam cross sectional area, this would correspond to a 1° SATA intensity in the focal plane of

$$I_{deg} = 255 \text{ mW/cm}^2.$$

With typical beam shapes this intensity would correspond more closely to the SPTA intensity:

$$I_{SPTA,deg} \approx 255 \text{ mW/cm}^2.$$

This in-water intensity is considerably below that of current doppler equipment; maximum general purpose duplex doppler system intensities have been reported as high as 2.3 to 4 W/cm^2 (16), and 1–2 W/cm^2 are not uncommon. If one rarely or never exceeded the 1° power, that would not be as stringent a restriction as the current FDA guidelines of Table 1, particularly for fetal imaging.

Conclusions

Since the calculations outlined in figures 5–7 are relatively easily implemented for on-screen labeling, it is appropriate to consider them for such, at least on systems in which the thermal index could exceed a specified level. Compared with other time averaged quantities such as I_{SPTA}, which may deviate from a linear relationship with potential temperature rise by several orders of magnitude, present calculations of thermal index probably are accurate for representing typical to worst case conditions within a factor of 2 or 4. It therefore would be worth providing on-screen labeling of the thermal index, with the understanding that refinements in the thermal index calculations may be required in the next few years. This should not be a terrible point of confusion if the underlying field parameters for the various ultrasound systems are readily available, as they always should be. At present, NCRP Committee 66 (9) is considering the following possible recommendations of figure 9.

a) If $W_o < W_{deg}$, do not withhold use of the ultrasound procedure because of concern about adverse effects from a thermal mechanism.
b) If W_o is greater than W_{deg}, use informed clinical judgement, considering anticipated benefits and risks.
c) In any case, do not utilize intensities or extend the duration of an examination beyond that necessary to obtain the desired diagnostic information.

Fig. 9. Proposed Recommendations

Similar indices of the potential for cavitation are not at all possible in the body at this time. However, I_{SPPA}, or some other temporal peak indicator such as maximum rarefactional pressure, p_r, will give adequate indication of relative proximity to cavitation thresholds. That will be particularly true if more complete acoustic data such as pulse duration is also easily available to the medical community in a format such as output specification sheets. Since availability of timely output information has continued to be a problem, it is also probably time to have one or more regional output calibration centers supported to make measurements and spot checks of medical ultrasound emissions and possibly performance. Currently there is no such service and publications of independently measured output emissions have been exceedingly rare, particularly in the United States.

There are several other areas in which we are lacking information, or progress has not been what one would desire. They are:

1. Redefinition of intensity as developed by Heyser
2. Evaluation of more sophisticated exposure parameters
3. Evaluation and approval of the AIUM/NEMA draft acoustic output measurement standard
4. Objective measures of the relations between diagnostic information and maximum available acoustic exposure variables.

Heyser (17, 18) provided a new definition of acoustic intensity equal to the complex acoustic pressure times the complex conjugate of the particle velocity, where the complex pressure is equal to the real pressure plus an imaginary component equal to the Hilbert transform of the pressure. The particle velocity is defined the same way (see also Ref. 18). This new intensity has considerable advantages, such as not showing O intensity at an antinode in a standing wave. It should be developed and utilized in medical ultrasound protection.

Regarding the AIUM/NEMA draft acoustic output measurement standard, its approval by NEMA has been delayed somewhat by current concerns over obtaining approval for marketing of Doppler ultrasound systems for fetal use. However, the modest set of output measurements required in the draft measurement standard are necessary, even if a small amount of information is required on-screen by a separate standard. The dual measurements at peak locations in water and at calculated *in situ* peaks are almost redundant.

As described above, the field of protection in diagnostic ultrasound is quite dynamic at this time. Assuming that a new standard or a set of standards is instituted, we will have to show great diligence in assuring that output information is obtained and distributed in a uniform and timely manner.

References

1. Ayme, E.J., Carstensen, E.L., Parker, K.G., Flynn, H.G. "Microbubble response to finite amplitude waveforms," in *Proceedings of the IEEE 1986 Ultrasound Symposium* (Institute of Electrical and Electron Engineers, New York) (1986).
2. AIUM. American Institute of Ultrasound in Medicine. "Bioeffects considerations for the safety of diagnostic ultrasound," J. Ultras. in Med., 7/9 (Sup.): S1–S38 (1988).
3. Holland, C.K. and Apfel, R.E. "An improved theory for prediction of microcavitation thresholds," IEEE Trans. Ultrasonics, Ferroelectrics and Frequency Control, **36,** 204–208 (1989).
4. Gross, D.R., Miller, D.L., and Williams, A.R. "A search for ultrasonic cavitation within the canine cardiovascular system," Ultras. in Med. & Biol., **11,** 85–97 (1985).
5. Williams, A.R., Delius, M., Miller, D.L. and Schwarze, W. "Investigation of cavitation in flowing media by lithotripter schockwaves both *in vitro* and *in vivo*," Ultras. in Med. & Biol., **15,** 53–60 (1989).
6. NCRP. National Council on Radiation Protection and Measurements. *Biological Effects of Ultrasound in Medicine*, NCRP Report No. 74 (National Council on Radiation Protection and Measurements, Bethesda, Maryland) (1983).
7. Carson, P.L., Rubin, J.M. and Chiang, E.H. "Fetal depth and ultrasound path lengths through overlying tissues," Ultras. in Med. & Biol. 15/7, 629–639 (1989).
8. Carson, P.L. "Constant-soft tissue distance model in pregnancies," in *Proceedings of the Second WFUMB Symposium on Safety and Standardization in Medical Ultrasound*, Airlie, Virginia, October 22–23 (in press).
9. NCRP. National Council on Radiation Protection and Measurements. SC-66, "Exposure criteria for medical ultrasound. Part 1. Criteria based on thermal mechanisms," Report being prepared for the NCRP by its Scientific Committee 66, expected finalization and publication in 1990.

10. AIUM/NEMA. American Institute of Ultrasound in Medicine/National Electrical Manufacturing Association. "Safety standard for diagnostic ultrasound equipment," J. Ultras. in Med. **2**/4 (Sup.): S1–S50 (1981).
11. Nyborg, W.L. "Solutions of the bio-heat transfer equation," Phys. Med. Biol., **33**, 785–792 (1988).
12. Nyborg, W.L. "NCRP-AIUM models for temperature calculations," in *Proceedings Second WFUMB Symposium on Safety and Standardization in Medical Ultrasound*, Airlie, Virginia, October 22–23, 1988 (in press).
13. Carstensen, E.L. "NCRP-AIUM uniform absorption models," in *Proceedings Second WFUMB Symposium on Safety and Standardization in Medical Ultrasound*, Airlie, Virginia, October 22–23, 1988 (in press).
14. Drewniak, J., Carnes, K.I. and Dunn, F. "Ultrasonic heating of fetal bone *in vitro*," J. Acoust. Soc. Am. (submitted).
15. Carstensen, E.L., Child, S.Z., Norton, S. and Nyborg, W.L. "Ultrasonic heating of the skull," (submitted).
16. Joint UK Health Departments Ultrasound Equipment Evaluation Project. "Evaluation of Acuson 128R ultrasound scanner," STD/87/18, (DHSS, Diagnostic Imaging Group, 14 Russell Sq. London WC1B, 5EP) (1987).
17. Heyser, D.H. Informal discussions. Allerton Conference, Allerton, IL. (1986).
18. Mann, J.A., Tichy, J. and Romano, A.S. "Instantaneous and time-averaged energy transfer in acoustic fields," J. Acoust. Soc. Am., **82** 17–30 (1987).

Scientific Session

Measurement and Dosimetry

Randall Caswell
Chairman

Radiation Protection Measurement Science: A Long Road Traversed but a Long Way to Go

Gail de Planque
Wayne M. Lowder
Harold L. Beck
Environmental Measurement Laboratory
U.S. Department of Energy
New York, NY

Introduction

The title of this paper conveys one of its essential messages. We will dwell mostly on some of the key historical developments that have brought the radiation protection community to its current fairly well-equipped state in terms of radiation measurement capabilities. Assessing the future is necessarily more speculative in nature, so that our brevity here is appropriately cautious. However, there is increasing evidence that there are major changes in the offing for radiation protection philosophy and methodology that may render obsolete some of our present measurement techniques. We will suggest some possible directions that future developments in measurement science may take in response to such changes, and hopefully encourage the type of research and development efforts that will be required to meet these new challenges. Thus, despite the many successes that have been achieved in building up our current arsenal of advanced measurement systems, it is unlikely that we will be able to rest on our laurels. We do have a long way to go.

In one respect, our title is overly ambitious. "Radiation protection measurement science" is too extensive a subject to adequately treat in the available space and time, and thus we have restricted our topic somewhat, hopefully without significant impact on the generic nature of our conclusions. Thus, we will consider radiation detector development in very broad categories, with only passing reference to some of the parallel advances in electrical and electronic engineering that have so radically improved our signal processing and interpretation capabilities. Because of our own particular experience and expertise, we focus specifically on the measurement of external radiation to develop and illustrate our main themes. Finally, our specific choices of historical highlights are intended to be illustrative rather than exhaustive, as are our suggestions for the future. We have chosen our references partly to provide the interested reader with the opportunity of obtaining more information and a broader perspective on the many aspects of this important subject that deserve fuller treatment. Particularly helpful in providing historical information and perspective are the papers by Kathren (1-4), Andrews (5), and Flakus (6).

The Early Years

Prior to the beginning of nuclear weapons development in the early 1940's, it would be difficult to specify any aspect of ionizing radiation measurement science that could be strongly identified with radiation protection practices, with the single major exception of film badge dosimetry. When the needs for measurement technology suddenly escalated in the 1940's, they were met by the application of techniques that had been developed over nearly half a century of basic research in many fields of science. To see how this happened, we present in Table 1 a brief chronology of key developments between the discovery of xrays in 1895 and the end of the last decade before the nuclear age was introduced with the Manhattan Project (a title that we use to cover all weapons development work done under the Manhattan Engineering District).

On that famous Friday afternoon in November 1895, Wilhelm Konrad Roentgen didn't realize that he was exposing a primitive scintillation detector (a screen of crystalline barium platino-cyanide) to xrays from an energized Crookes tube. But the flourescence from that screen, and subsequent images on photographic plates (Figure 1), alerted him and the world at large to an exciting new phenomenon. Within a few months, Henri Becquerel noted the blackening of photographic plates situated

TABLE 1—*Radiation Detector Development: Historical Highlights (1895–1940)*

1895:	DISCOVERY OF X RAYS USING A FLUORESCENT SCREEN [ROENTGEN]
1896:	DISCOVERY OF RADIOACTIVITY USING A PHOTOGRAPHIC PLATE [BECQUEREL]
1896:	ELECTROSCOPES APPLIED TO RADIATION DETECTION [BECQUEREL *ET AL.*]
1897:	FIRST CALORIMETER, USING AN AIR THERMOMETER [DORN]
1899:	OPERATING CHARACTERISTICS OF IONIZATION CHAMBERS DEFINED [THOMSON]
1899:	DISCOVERY OF PENETRATING ENVIRONMENTAL RADIATION USING ELECTROSCOPES [ELSTER, GERTEL, WILSON]
1902:	FIRST CHEMICAL DOSIMETER, THE CHROMORADIOMETER, BASED ON THE DISCOLORATION OF A SOLID-FUSED SECRET MIXTURE [HOLZKNECHT]
1903:	DEVELOPMENT OF COLORIMETRIC CHEMICAL DOSIMETER [SABOURAUD AND NOIRE]
1903:	DEVELOPMENT OF DIRECT-READING SCINTILLATOR (ZnS SCREEN), THE SPINTHARISCOPE [CROOKES]
1903:	FIRST USE OF FILM TO ASSESS MEDICAL X-RAY EXPOSURE
1906:	FIRST DESCRIPTION OF PARALLEL-PLATE AND CYLINDRICAL IONIZATION CHAMBERS [RUTHERFORD]
1908:	DEVELOPMENT OF CYLINDRICAL GAS COUNTER, THE GEIGER TUBE [RUTHERFORD AND GEIGER]
1912:	DISCOVERY OF COSMIC RADIATION USING BALLOON-BORNE IONIZATION CHAMBER SYSTEMS [HESS]
1921:	FIRST USE OF FILM FOR ROUTINE PERSONNEL MONITORING OF X-RAY EXPOSURE
1922–30:	VARIOUS VERSIONS OF CONDENSER ION CHAMBER DEVELOPED AND COMMERCIALIZED [DUANE, SOLOMON, GLASSER, FRICKE, SEITZ, VICTOREEN]
1926:	INTRODUCTION OF FIRST X-RAY FILM BADGE DOSIMETER [QUIMBY]
1928:	DEVELOPMENT OF GEIGER-MULLER COUNTER
1928:	FIRST MEASUREMENTS USING PROPORTIONAL COUNTING TECHNIQUES [GEIGER AND KLEMPERER]
1929–30:	FIRST PORTABLE IONIZATION CHAMBER SYSTEM FOR RADIATION PROTECTION APPLICATIONS [TAYLOR]
EARLY 1930s:	DEVELOPMENT OF PRESSURIZED ARGON ION CHAMBER SYSTEMS FOR COSMIC RADIATION MEASUREMENTS [MILLIKAN, NEHER]
LATE 1930s:	INTRODUCTION OF FIRST COMMERCIAL PORTABLE ION CHAMBER AND GEIGER-MULLER COUNTER [VICTOREEN]
1940:	INTRODUCTION OF COMMERCIAL POCKET ION CHAMBER FOR PERSONNEL DOSIMETRY [VICTOREEN]
1940:	DEVELOPMENT OF GRIDDED IONIZATION CHAMBER [FRISCH]

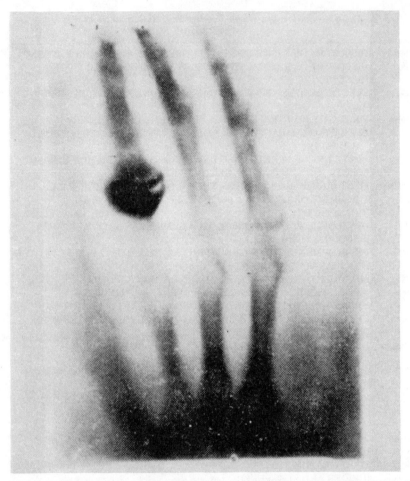

Fig. 1. Early radiograph taken by W.C. Roentgen. (Photograph courtesy of Ron Kathren.)

near samples of uranium-bearing ore. He and others also quickly noted that electroscopes were sensitive to the presence of these new radiations. Thus, the discovery of ionizing radiation and radioactivity went hand in hand with the discovery and application of radiation detection systems.

These techniques for radiation measurement were quickly applied at the forefront of research in atomic and nuclear physics. Within a few years, the basic operating characteristics of ionization chambers were well understood, and the Curies used such chambers to make some of their greatest discoveries. As Table 1 indicates, other techniques were soon discovered to detect and measure ionizing radiation. Of particular interest was the application of film techniques to the determination of the x-ray exposures of patients and radiologists, beginning in 1903 as the medical community became gradually aware of some of the not-so-benign effects of such exposure. As Kathren (4) has pointed out, the use of film strips at this time to ascertain qualitatively the degree of patient exposure to xrays may have been the first radiation protection measurements in history. For the next several decades, the main challenges in radiation protection were in the medical field. Kathren (2) has given a fascinating summary of the historical development of x-ray protection during this period.

Prior to World War I, there were important advances in radiation measurement science associated with basic studies of radioactivity and atomic physics. Rutherford's group at the Cavendish Laboratory, Cambridge, was associated with many of these advances, which included refinements in ionization chamber capabilities, initial development of the Geiger tube, and the application of scintillation techniques to the counting of individual particles. Ionization chamber technology was applied to the study of environmental radiation as early as 1899, and systems flown in balloons in 1910–12 provided the data that resulted in the discovery of cosmic rays.

In the '20s and '30s, cosmic-ray studies were at the forefront of nuclear physics research and stimulated the development of improved electrometers and the pressurized argon-filled steel-walled ionization chamber. These systems were flown in aircraft and on balloons to altitudes in excess of 30 km and taken on ships around the world to map the cosmic-ray field (*e.g.*, 7).

Despite some absolute calibration problems (8), these chamber systems set the stage for later developments in this technology. However, no simple portable instruments existed for more earth-bound measurements, though Taylor (9) took the first steps in this direction in 1929 with a somewhat bulky instrument using three aluminum-walled air-filled ion chambers.

Two other important developments in the evolution of ion chamber technology during this period were the introduction of condenser R-meters in the late 1920's, which soon became commercialized for use in practical x-ray dosimetry for diagnostic and therapeutic applications, and the pocket ion chamber introduced in 1940, which soon became a valuable real-time personnel monitor. The condenser chamber became central to our ideas of not only what could be measured but also what should be measured. It is no accident that the definitions of the quantity,

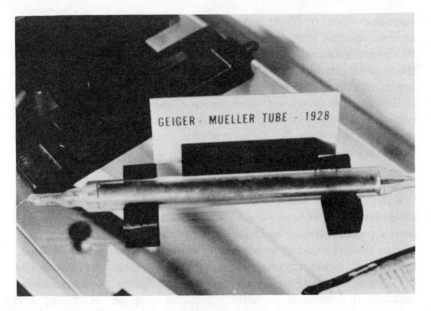

Fig. 2. Early example of a G-M tube. (Photograph courtesy of Ron Kathren.)

exposure, and its unit, the roentgen, were closely related to the response of these ion chambers.

Another gas-filled chamber became widely utilized in the 1930's, the Geiger-Muller counter. Although the basic principle of operation was known and had been utilized since 1908, it was only after important improvements were introduced between 1928 and 1937 that practical instruments could be designed for radiation protection applications. These improvements included tube designs and gas fillings that allowed large volumes, fast responses, and elimination of expensive electrical circuits. An early example of a G-M tube is shown in Figure 2. Such detectors were commonly used in many nuclear and particle physics studies, and accompanied ion chambers on cosmic-ray balloon flights.

Prior to 1940, there were many important advances in signal-processing capabilities, notably the development of stable linear amplifiers, highly sensitive vacuum-tube electrometers, and vastly improved scaling and counting circuits. Both in this area and with respect to the radiation detection systems mentioned above, the stage was set for the literally as well as figuratively explosive era of the 1940's.

The Modern Era

It is not difficult to regard the state of radiation measurement technology in 1940 as being rather primitive, especially relative to our present capabilities or even those of, say, 1960. But the rapid advances that were made during the 1940's, stimulated by the demands of the Manhattan Project, can only be partly explained by the ingenuity and creativity of the pioneers in the development of radiation protection instrumentation. For the most part, it was actually the successful adaptation and modification of available technologies that produced what turned out to be a highly effective response to these new demands. By 1945, the practical personnel monitors, *i.e.*, film badges and pocket ionization chambers, and area monitors, *i.e.*, ionization chambers and GM counters, were still recognizable variants of detectors that had found previous applications in basic physics research and in diagnostic radiology and medical physics.

In making this point, we are not disagreeing with Kathren (1, 2), who in his historical surveys has rightly emphasized the many achievements of the radiation protection programs of the Manhattan Project as being seminal to the future of radiation protection measurement science. But it is important to recognize the elements of continuity through this period of dramatic changes.

The nature and extent of the advances made during the 1940's can only be hinted at in the context of this presentation. Kathren (4) has pointed out that 16 different types of survey meter were produced during the Manhattan Project, including the well-known Juno and Cutie Pie. Improved versions of some of these instruments are still in use. A compensated ion chamber system, called the Chang and Eng after a famous pair of Siamese twins, was developed for neutron monitoring (see Figure 3), but this was soon superseded by the moderated BF_3 system. Although the basic principles of proportional counting had been known since early in the century, practical proportional counters for alpha monitoring and spectrometry were a product of this period. Film badge design and performance were considerably improved, and methods were developed for more accurate dosimetry. Personnel dosimetry also benefitted from more rugged, reliable, and sensitive pocket ionization chambers. The first steps were taken to develop practical scintillation detectors, but the important developments in this technology and in pulse height analyzers that provided important spectrometric capabilities were to take place in the 1950's. However, the key point here is that by 1950 there was a nuclear industry in place that included a large population of radiation workers and there were radiation measurement technologies available that could adequately address the radiation protection needs of this population. These technological advances were also of course relevant to the improved protection of medical practitioners and patients.

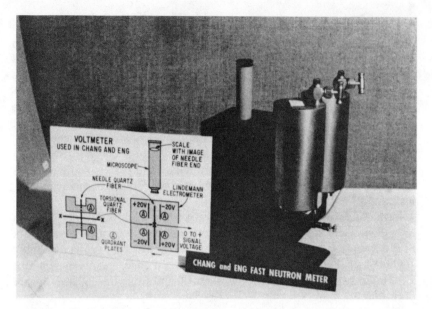

Fig. 3. Chang and Eng neutron monitor.

In the 1950's, a new dimension was added to the demands on radiation protection measurement science, namely, increasing awareness of the potential for radiation exposure of the general public from nuclear defense and nuclear energy activities. Concern about such exposure on a large scale first arose when it became evident that nuclear weapons tests were producing global fallout deposition that resulted in exposures of large populations to both external radiation and radiation from internally-deposited nuclides. Later, as nuclear reactors and other nuclear facilities were built, similar concerns stimulated efforts to document any possible general population exposure pathways due to their routine operation and to assess the potential consequences of any accident resulting in radionuclide releases to the environment. In short, a third category of radiation measurement was added to already-existing area and personnel monitoring, namely, environmental monitoring. Since environmental measurements usually address situations where exposures of members of the general public might take place, the need arose for risk assessment as a goal in the interpretation of these measurements. Ideally, whether dealing with routine or emergency situations, one would like to have reasonable estimates of actual doses to real people. And today perhaps we're learning that the need for a reasonably accurate dosimetry is much the same inside the plant as outside. Fortunately, because of the advances made in radiation detection instrumentation during the 1940's, the tools needed to begin to address the new demands on radiation measurment technology were either at hand or were rapidly developed.

In this context, we can only mention here a few of the highlights of radiation detector development since 1950 that are relevant to radiation protection applications in both the occupational and off-site environments. Of immense importance was the gradual introduction of solid-state technology into signal processing and radiation detection applications. In the 1960's, thermoluminescence dosimeters became widely utilized as personnel monitors and in the 1970's they were shown to be useful environmental monitors(10). Silicon and germanium detectors together with vastly improved multichannel analyzers provided high-resolution spectrometric capabilities important for both dosimetric and diagnostic purposes in area and environmental surveys. It should be noted once again that this solid-state revolution generally derived from research areas far removed from radiation protection. However, in some cases, notably in the development of thermoluminescence detectors, the potential radiation protection applications were a driving force in the developmental effort. Here are the true beginnings of identifiable research of a basic nature that can be assigned to radiation protection measurement science.

Our laboratory's contributions to this progress were, as its name implies, primarily in the environmental measurement category, though we have also been involved in addressing problems more commonly associated with area and personnel monitoring. For example, we modified the pressurized argon ionization chamber technology that had been developed in the 1930's to meet more rigorous requirements on ruggedness, sensitivity, and energy response, and produced several versions of the same basic design that have found world-wide application (11, 12). We developed the methodology for *in-situ* gamma spectrometry using scintillation and solid-state detectors to infer radiation exposure rates and radionuclide concentrations (*e.g.*, 13). We also evaluated detectors and procedures for the environmental applications of thermoluminescence dosimetry (*e.g.*, 10, 14). And we've also been able to make some contribution to state-of-the-art neutron measurement techniques in area and environmental monitoring using Bonner sphere spectrometry (*e.g.*, 15-17), and to the measurement of beta radiation fields using a thin-walled dual concentric ionization chamber system (18). The important point to note here is the fact that each of these advances involved to varying degrees the adaptation of an existing radiation detection technology to particular problems in radiation protection. The innovative

TABLE 2—*Radiation Measurement Techniques: 1990*

Type	Radiations Detected	Quantities "Measured"	Monitoring Applications
Gas Ionization	α, β, γ, n	φ, K, D	A, P, E
Solid Ionization	α, β, γ, n	φ, K, D	A, E
Scintillation	α, β, γ, n	φ, K, D	A, E
TLD	α, β, γ, n	K, D	P, E
Film	β, γ, n	K, D	P
Nuclear Track	α, n		P, E
Bubble	n		A, P
Activation	n		A, P

Note: φ = Flux density
K = Kerma
D = Absorbed dose
A, P and E refer to—Area, personnel and environmental monitoring, respectively

aspects resided primarily in the new methods of data analysis and interpretation developed to optimally achieve the desired goals, and in some cases in instrument design characteristics that improved the quality and interpretability of the data, rather than in any major improvements in detector or signal processing technology. This situation has been common in the development of the present arsenal of weapons for radiation protection measurements.

The Arsenal Today

Having indicated some of the highlights in the development of radiation detection technologies that have been applied to radiation protection problems, it is appropriate to summarize the current status. In Table 2, we list those types of detectors that are widely used for area, personnel, and environmental monitoring and measurement, the radiations that are normally measured by these generic detector types, and an indication of whether the response is functionally related to the flux density of incident particles and/or to the dose or kerma in the detector medium. This last property may have important implications for the future, since biological damage appears to be more closely related to events occurring along the tracks of the individual particles than to average energy deposition per gram of material. Also indicated are the monitoring applications for which the various detector types have found extensive use.

Two terms used in Table 2 perhaps deserve some explanation. By "solid ionization" detectors, we are referring to that array of solid-state detectors that operate by means of the separation and collection of ions generated in the solid medium, *e.g.*, silicon and germanium diodes. The "bubble" detector category refers to superheated liquid or plastic media where radiation-generated energy absorption events above a certain energy threshold produce vapor bubbles that can be counted.

There are a great number of different types of detectors that are available within many of the broad generic categories of Table 2. Moreover, today the existing technologies are rapidly changing, due in particular to advances in microelectronics and materials science that have produced smaller, cheaper, more rugged, and more reliable detectors and instrument packages. Very often the detector itself determines the size and weight of the instrument. More importantly, such instruments are becoming "smarter," *i.e.*, capable of faster response or more rapid and more extensive data processing. The instruments themselves can accomplish tasks that previously required external mechanisms. Bellian (19) has emphasized the point that health physics instrumentation is currently undergoing a major transformation from discrete-component and integrated-circuit to microprocessor base. All of these changes significantly improve our measurement capabilities.

Despite this vast arsenal of radiation detection systems and the great improvements in instrument intelligence now taking place, all is not well. We still cannot adequately characterize some complex radiation fields, notably mixed beta-gamma and neutron-gamma fields. We don't have an adequate personnel monitor for neutrons of intermediate energies, an energy region of considerable dosimetric importance under some exposure conditions at nuclear reactors. There are still problems associated with determining the relevant doses to individuals exposed to beta and low-energy photon fields. Lucas (20) makes the straightforward statement that "no satisfactory instrumentation has been developed for the accurate measurement of beta ray and neutron dose equivalents." Griffith (21) is somewhat more positive, asserting that "compared with photon dosimetry, neutron and beta measurements...remain partially unsolved dosimetry problems." In any case, much remains to be done in these areas.

To complicate matters further, there is the much discussed question of the appropriate quality factors to apply to absorbed dose estimates in order to derive dose equivalents, especially for neutrons. The NCRP (22) has addressed this issue in a preliminary way, but there are many outstanding questions. However, in the long run, they may prove to be irrelevant, as radiation protection philosophy may be gradually moving away from the basic concept of dose equivalent as the relevant limiting quantity. But that raises other questions ... which leads us to briefly consider what the future might bring.

Prospects for the Future

As pointed out, one can reasonably expect continued refinements in instrument design and operational characteristics. Modest improvements in radiation detector technology will certainly continue. For example, Umbarger (23) has indicated likely developments for improved semiconductor detectors, *e.g.*, GaAs, CdTe, and HgI_2. Along with Griffith (21), we can feel some optimism that the problems associated with beta and neutron measurements can be at least partially overcome. But the next revolution may well be arising with the likelihood of major developments in "microdosimetric" detector systems.

Microdosimetry has been an active field of research for about three decades. Various types of models of biological effects resulting from processes initiated by energy absorption events on the microscopic scale have been developed, but progress has been slow because of the extreme complexity of the operative processes and our very limited knowledge of what may be the relevant parameters (*e.g.*, electron energy loss mechanisms at low energies). However, from the standpoint of measurement science, it is becoming clear that absorbed dose is a physical quantity that is only very indirectly related to health risk; we probably should be looking at such things as the number, size, and density of individual energy absorption events along particle tracks if we desire to measure something that might be a more valid surrogate for risk than dose equivalent. A number of approaches are being tried to accomplish this. The most advanced is the low pressure tissue-equivalent proportional counter, first developed by Rossi and Rosenzweig (24) and now widely utilized to measure event-size distributions in simulated spherical volumes (1 g/cm^3) with diameters ranging from a few micrometers down to a few nanometers, *i.e.*, from cellular to DNA-strand dimensions (*e.g.*, 25–27). Practical versions of this type of detector are currently being evaluated (28, 29). Another approach is the above-mentioned bubble-type detector, two versions of which are presently being commercialized for personnel dosimetry, one involving the production of drops in a supersaturated soft gel (30) and the other involving the formation of bubbles in an elastic polymer (31). Hamm *et al* (32), at ORNL, are developing an optical ionization chamber for digital characterization of primary and secondary charged particle track structure. A microdosimeter based on the production of soft upsets in dynamic random access memory chips is an attractive idea that is being investigated in several laboratories. The common theme to all of these approaches is the recording of a possible initiating event for any biological damage process. This seems to be a direction that the future evolution of radiation protection measurement science should take if the current trend toward a risk-based system of radiation protection continues.

However, it should be noted that, even if more relevant quantities could be adequately measured, one is not usually measuring them at the location or in the medium of interest, namely, at the point of potential biological damage within the body. The transformation of radiation detector response to human receptor response can sometimes be made by the use of calculated conversion factors that apply under specific irradiation conditions and source-detector geometries, as for example are given by ICRP (33) and ICRU (34). An alternate (and clearly related) approach is the application of measurement technology to the determination of the properties of the radiation field (*i.e.*, specifying the irradiation conditions), and then making use of our rapidly developing computational capabilities to infer the desired quantities inside a human receptor. Many such capabilities can be built into an instrument using advanced microprocessor technology, and the calculations directly incorporated into the measurement process. Changes in algorithms or recommended conversion factors can be readily accomodated, greatly increasing the ability of these instruments to adjust to and meet future demands. In any case, as fundamental research in radiation physics, chemistry, and biology provides further indications of what should be measured and as investigations of new radiation detector systems indicate what quantities might be practically measured, we will also need to refine our calculational capabilities to develop a new generation of conversion factors essential to the proper interpretation of radiation protection measurements.

These considerations certainly suggest some of the important needs for future research and development, as well as our choice of title. There's still much to be done. We fully expect that the NCRP will continue to play a vital role in helping to define future directions for this research, assessing the state-of-the-art in various areas of measurement science, and providing guidance on appropriate conversion factors for the interpretation of measurements. A new NCRP report (35) addresses the problem that may ultimately prove to be a key driving force for a new philosophy and a new kind of measurement science, namely, the exposure of participants in space missions to a harsh radiation environment of unprecedented magnitude. Bouville and Lowder (36) have alluded to some of the implications of the gradual increase in the exposure of the human population to cosmic radiation due to activities in space, notably at least the modification of our current concept of dose equivalent when the need to assess risk to real people under unavoidable or only partially controllable irradiation conditions becomes paramount.

In space, this assessment would be the central task for radiation protection measurements. We in turn can only allude to some of the promising directions for measurement science research and development that might be responsive to this problem and to likely trends in radiation protection philosophy. However, we do have more confidence in the notion that the greatest challenges to radiation protection measurement technology may yet lie ahead.

References

1. Kathren, R.L. "Before transistors, IC's and all those other good things: the first fifty years of radiation monitoring instrumentation," pages 73-93 in *Health Physics: A Backward Glance*, Kathren, R.L. and Ziemer, P.L. Eds. (Pergamon Press, Oxford) (1980).
2. Kathren, R.L. "Historical development of radiation measurement and protection," pages 13-52 in *CRC Handbook of Radiation Measurement and Protection, Section A, Part 2*, Brodsky, A.B., Ed. (CRC Press, Boca Raton, Florida) (1982).
3. Kathren, R.L. "Instrumentation for monitoring and field use," Ch. 18 in *Radioactivity and Health: A History*, Stannard, J.N., Ed., DOE/RL/01830-T59 (Pacific Northwest Laboratory, Richland, Washington) (1988).
4. Kathren, R.L. "Health physics instruments yesterday: some notes on the evolution of health physics instruments," pages 1-9 in *Proceedings 22nd Midyear Topical Meeting on Instrumentation*, San Antonio, Texas, 1988 (Health Physics Society, San Antonio, Texas) (1988).
5. Andrews, H.L. "Laboratory measuring instruments," Ch. 17 in *Radioactivity and Health: A History*, Stannard, J.N., Ed., DOE/RL/01830-T59 (Pacific Northwest Laboratory, Richland, Washington) (1988).
6. Flakus, F.N. "Detecting and measuring ionizing radiation - a short history," IAEA Bulletin **23,** No. 4, 31 (1981).
7. Neher, H.V. "Recent data on geomagnetic effects," pages 243-314 in *Progress in Cosmic Ray Physics*, Wilson, J.G., Ed. (North Holland, Amsterdam) (1952).
8. NCRP. National Council on Radiation Protection and Measurements. *Exposure of the Population of the United States and Canada from Natural Background Radiation*, NCRP Report No. 94 (National Council on Radiation Protection and Measurements, Bethesda, Maryland) (1987).
9. Taylor, L.S. "An early portable survey meter," Health Phys. **13,** 1347-8 (1967).
10. de Planque, G. and Gesell, T.F. "Thermoluminescence dosimetry - environmental applications," Int. J. Appl. Radiat. Isot. **33,** 1015-1034 (1982).
11. Beck, H.L., Lowder, W.M. and McLaughlin, J.E. "*In situ* external environmental gamma-ray measurements utilizing Ge(Li) and NaI(Tl) spectrometry and pressurized ionization chambers," pages 499-513 in *Rapid Meth-*

ods for Measuring Radioactivity in the Environment, IAEA-SM-148/2 (International Atomic Energy Agency, Vienna) (1971).
12. Latner, N., Miller, K., Watnick, S. and Graveson, R.T. "SPICER, a sensitive radiation survey meter," Health Phys. **44,** 379-386 (1983).
13. Beck, H.L., DeCampo, J. and Gogolak, C. "*In situ* Ge(Li) and NaI(T1) gamma-ray spectrometry," USAEC Report HASL-258 (1972).
14. de Planque, G. and Gesell, T.F. "Environmental measurements with thermoluminescence dosimeters - trends and issues," Radiation Prot. Dosim. **17,** 193-200 (1986).
15. Hajnal, F., McLaughlin, J.E., Weinstein, M.S. and O'Brien, K. "1970 sea-level cosmic-ray neutron measurements," USAEC Report HASL-241 (1971).
16. Hajnal, F. and Griffith, R.V. "Least squares unfolding and error analysis of Bonner sphere neutron measurements: application to PWR and "Little Boy" critical assembly measurements," pages 415-425 in *Proceedings Fifth Symposium on Neutron Dosimetry* Report EUR 9762, *Vol. 1*, Schraube, H., Burger, G. and Booz, J., Eds. (Commission of the European Communities, Luxembourg) (1985).
17. Hajnal, F. "1986 cosmic-ray neutron measurements," EML Report (in press).
18. Hajnal, F. and McLaughlin, J.E. "Dosimetry of mixed beta and gamma radiation in work areas," pages 103-114 in *Proceedings IAEA Symposium on Advances in Radiation Protection Monitoring*, Stockholm, Sweden (International Atomic Energy Agency, Vienna) (1979).
19. Bellian, J.G. "Health physics instruments today," pages 10-13 in *Proceedings 22nd Midyear Topical Meeting on Instrumentation*, San Antonio, Texas, 1988 (Health Physics Society, San Antonio, Texas) (1988).
20. Lucas, A.C. "The future of radiological instrumentation," Health Phys. **55,** 191 (1988).
21. Griffith, R.V. "The next decade in external dosimetry," Health Phys. **55,** 177-189 (1988).
22. NCRP. National Council on Radiation Protection and Measurements. *Recommendations on Limits for Exposure to Ionizing Radiation*, NCRP Report No. 91 (National Council on Radiation Protection and Measurements, Bethesda, Maryland) (1987).
23. Umbarger, C.J. "Health physics instruments tomorrow," pages 21-26 in *Proceedings 22nd Midyear Topical Meeting on Instrumentation*, San Antonio, Texas, 1988 (Health Physics Society, San Antonio, Texas) (1988).
24. Rossi, H. and Rosenzweig, W. "A device for the measurement of dose as a function of specific ionization," Radiology **64,** 404 (1955).
25. Menzel, H.G. "Practical implementation of microdosimetric counters in radiation protection," pages 287-307 in *Proceedings Fifth Symposium on Neutron Dosimetry*, Report EUR 9762, *Vol. 1*, Schraube, H., Burger, G. and Booz, J., Eds. (Commission of the European Communities, Luxembourg) (1985).
26. Menzel, H.G. "Application of low pressure proportional counters to radiation protection," in *Proceedings Tenth Symposium on Microdosimetry*, Rome, 1989 (in press).

27. Goldhagen, P., Marino, S.A. and Kliauga, P. "Variance covariance measurements of y_D for 14 MeV neutrons in spherical sites with a wide range of sizes," in *Proceedings Tenth Symposium on Microdosimetry*, Rome, 1989 (in press).
28. Dietze, G., Booz, J., Edwards, A.A., Guldbakke, S., Kluge, H., Leroux, J.B., Lindborg, L., Menzel, H.G., Nguyen, V.D., Schmitz, Th. and Schuhmacher, H. "Intercomparison of dose equivalent meters based on microdosimetric techniques," Radiat. Prot. Dosim. **23,** 227 (1988).
29. Dietze, G., Edwards, A.A., Guldbakke, S., Kluge, H., Leroux, J.B., Lindborg, L., Menzel, H.G., Nguyen, V.D., Schmitz, Th. and Schuhmacher, H. "Investigation of radiation protection instruments based on tissue-equivalent proportional counters: results of a Eurados intercomparison," Report EUR 11867 (Commission of the European Communities, Luxembourg) (1988).
30. Apfel, R.E. and Roy, S.C. "Investigations on the applicability of superheated drop detectors in neutron dosimetry," Nucl. Instr. Meth. **219,** 582 (1984).
31. Ing, H. and Birnboim, H.C. "Bubble-damage polymer detectors for neutron dosimetry," pages 883–894 in *Proceedings Fifth Symposium on Neutron Dosimetry*, Report EUR 9762, *Vol. 2*, Schraube, H., Burger, G. and Booz, J., Eds. (Commission of the European Communities, Luxembourg) (1985).
32. Hamm, R.N., Turner, J.E., Wright, H.A., Hunter, S.R., Hurst, G.S. and Gibson, W.A. "Analysis of data from an optical ionizing-radiation-track detector—an optical electron detection technique for the observation of ionizing radiation particle tracks in a gas," in *Proceedings Tenth Symposium on Microdosimetry*, Rome (in press).
33. ICRP. International Commission on Radiological Protection. *Data for Use in Protection Against External Radiation*, ICRP Publication No. 51 (Annals of the ICRP **17,** No. 2/3) (Pergamon Press, Oxford) (1987).
34. ICRU. International Commission on Radiation Units and Measurements. *Determination of Dose Equivalents Resulting from External Radiation Sources—Part 2*, ICRU Report No. 43 (International Commission on Radiation Units and Measurements, Bethesda, Maryland) (1988).
35. NCRP. National Council on Radiation Protection and Measurements. *Guidance on Radiation Received in Space Activities*, NCRP Report No. 98 (National Council on Radiation Protection and Measurements, Bethesda, Maryland) (1989).
36. Bouville, A. and Lowder, W.M. "Human population exposure to cosmic radiation," Radiat. Prot. Dosim. **24,** 293 (1988).

Discussion

YOKOZUMA, (FDA): I would warn about the overuse of the term receptor because receptor has plenty of different meanings in different areas of biology and biochemistry.

GAIL DE PLANQUE: That may be a good point and from what we heard earlier from Dr. Lushbaugh, he may have more material at the next NCRP meeting.

CHARLIE MEINHOLD: I think you hit a very important point when you talked about the idea that the computer as part of the instrument may allow us to back up in a sense. One of our difficulties is that we've gotten better radiation protection instruments. We've lost more and more information about the fields at which the people are exposed. And I think this question of perhaps someday returning to a field quantity such as fluence and let the computer do what ever you like with it and still retain the original information may well be the way we need to be thinking for the future.

GAIL DE PLANQUE: Yes, as you noticed when I showed the slide of what's in our current arsenal, I indicated that those instruments measure fluence, kerma, and dose. And if you can measure these basic quantities and use as you are saying computational techniques, we may have a better match between what we can really measure and what we want to know.

AL TSCHAECHE (WINCO): Gail, we now have a term called effective dose equivalent. Are you or do you know of anybody who's doing any work, particularly with neutron fields, where we have a single, let's say, TLD system on the chest, unfold the information from that single TLD, do the calculation for each organ dose, and then add everything back up again?

GAIL DE PLANQUE: There have been a number of published papers dealing with the relationship of individual dosimeter readings to various dose equivalent quantities. The key point is that one has to have other information besides the dosimeter reading to make the conversion.

AL TSCHAECHE (WINCO): I would like to make a comment that there has been a development in the ICRU attempting to use quantities which relate to the effective dose equivalent. And these are being pursued very actively in Europe as are the tissue equivalent proportional counter technology. And not much is being done in the United States in either of these fields. I think this is an opportunity that we should maybe look into.

RON KATHREN: Thanks to Gail for the kind words. I just want to point out that there is one little footnote to this that should be mentioned, perhaps, and that is the use of biological dosimeters and even people themselves. This is often done following accidental exposures. It would be one hope for the nineties perhaps that this could be extended to lower level measurements and actually use people themselves as their own dosimeters. That may be pie in the sky but perhaps that's what we should be shooting for.

GAIL DE PLANQUE: That's an interesting point, and people have dabbled in this area over the years, looking at TL from teeth, for example, and watch jewels. But there's not an awful lot of work going on in that field and perhaps more effort should be put in that area of research.

KUMAZAWA (Japan): In your lecture, we cannot find reference to a glass dosimeter. Nowadays we can use laser light to stimulate a glass dosimeter. Before we only used ultraviolet light. How about the glass dosimeter?

GAIL DE PLANQUE: That's very true. And there's another interesting application of laser technology that started out awhile ago using laser readers for thermoluminescence dosimeters which would allow one to pick out just small parts of that detector; use something like a wedge filter and get a basic spectrometer. Not a lot of money has gone into this research but that in connection with the glass also has possibilities.

Radiation Dosimetry—Past and Present

Kenneth R. Kase
University of Massachusetts Medical Center
Worcester, MA

Abstract

The development of the dosimetry of ionizing radiation began with the discovery of x rays in 1895. From very early on, photographic film and air-ionization detectors have been used to measure the presence of ionizing radiation and these have been supplemented by additional types of detectors over the years. Beginning with the concept of cavity ionization and its relationship to energy deposited in air or a surrounding medium, much effort has been devoted to relating the response of instruments to the energy absorbed by matter from an external ionizing radiation field. Similarly, significant effort has been devoted to developing calculational models of biological systems and energy deposition distributions so that radiation dose from internally deposited radioactivity can be estimated. This requires a knowledge of the physical decay characteristics of the radionuclide as well as its biological behavior. In relating radiation field quantities to dose, interfaces between materials of different densities and composition, and other situations that lead to the lack of charged particle equilibrium have presented a challenge that has been effectively, but not completely, addressed by advanced Monte Carlo simulation techniques. Additionally, the concept of Quality Factor has been advanced to relate the energy absorbed by living tissue to the resulting biological damage. This has led to the development and refinement of the quantities Dose Equivalent and Effective Dose Equivalent. These important developments in radiation dosimetry have brought us to the present sophisticated status of both physical and biological dosimetry. Challenges still remain in the areas of charged particle dosimetry, doses at interfaces, dosimetry of internally incorporated emitters and elsewhere. Present and future needs in radiation dosimetry include addressing these challenges by developing more accurate dose calculational and biological models, and the experimental techniques necessary to verify them.

Before the "Roentgen"

Very soon following the discovery of x rays it was recognized that these new rays could be detected by the ionization that they created in air. Also, beginning in 1896, there were reports of damage to human tissues ascribed to x rays. During the first decade after Roentgen's discovery the field of radiation dosimetry was chaotic, but a number of techniques were attempted to measure the intensity of x ray fields and personal exposure to x rays. These included measurement of ionization in air and darkening of photographic plates, as well as temperature and color changes. However, these methods were all highly subjective and there was no correlation among them of the relationship of the measured quantity to the radiation field intensity.

The search for a dosimetry system was driven by the use of x rays for therapeutic purposes (1), and in 1908 a direction began to emerge with a proposal by Villard (2) that focused measurement efforts on the ionization technique. The proposal was that the unit for the quantity of x rays be "the electrical charge in electrostatic units produced in air which would be liberated per cubic centimeter of air under normal conditions of temperature and pressure." Measurement difficulties persisted, however, and the idea was ignored until 1910 when B. Szillard described the first small (1 cm^3) ionization chamber (1) and suggested using 10^{12} ions/cm^3 as the fundamental unit. Unfortunately, Szillard failed to realize that the ionization in his chamber was, in large part, produced by secondary electrons originating in the walls, and thus it could not reproduce his unit.

Over the next 10 years this idea was further developed, measurement techniques were improved and several units relating to ionization in air were proposed. In 1914 Duane recognized the reason for the failure of Szillard's technique and constructed a large ionization chamber that avoided the wall effect (1). The method he described was to measure the radiation field in "E" units (1 esu/cm^3) in the large chamber and then transfer the measurement to a small ionization chamber for use. The fundamental process has changed little since then for the dosimetry of low energy x rays.

The first suggestion of a unit designated "R" was made by Solomon in 1921 (1). This unit related ionization intensity to the radiation field produced by radium. It was defined as "the intensity of a roentgen radiation producing the same ionization per second as 1 g of radium placed 2 cm from the ionization chamber and filtered by 0.5 mm of platinum. A number of other similar units were proposed to quantify the ionization in air produced by x rays. Eventually, the ideas of Villard and Duane would be refined into the definition of the unit "roentgen" for a later-defined quantity "exposure".

Meanwhile, Szillard had recognized in 1914 that the charge per unit volume as defined by Villard was equivalent to the energy absorbed per unit volume of air because it must be proportional to the energy imparted to air by the radiation. At the same time Christen (3) discussed the concept of energy absorbed per unit volume and related that to a "dose" of x rays. In its early formulation Christen described "dose" as the product of energy fluence and an attenuation coefficient. His method did not distinguish among the various energy transfer interactions, or between energy transfer and energy absorption, and it assumed that the energy transfer from a photon to the absorbing material took place at the point of interaction, *i.e.* transport of electrons was not recognized. He did recognize the importance of beam "quality" in the absorption of x rays and suggested the "half-value layer" (HVL) as a measure of the quality. Another insight that Christen expressed was that the absorbed energy per unit mass was about the same for air and water. These ideas began to address the serious problem of a quantity that could be directly related to the effects of radiation on the absorbing material. Over the next decade they were refined and by 1924 it was recognized that:

1. The biological dose (related to biological effects) could not be measured exactly because the effects depend on many biological factors.
2. The radiation quality is not determined by the measurements so far proposed for determining radiation intensity.
3. The ratio of readings in the small ion chamber to those of the large ion chamber (standard) changed with beam quality.
4. The design and construction of the small ion chambers resulted in faulty measurements in many cases.

Attempts to relate radiation field quantities to energy absorption and biological effects continued. Several years before the official adoption of the roentgen unit Meyer and Glasser (4) related some observed effects of x-ray exposure to the R unit as defined by Behnken. They reported that, "The roentgen ray quantities, measured in R units, which are nec-

TABLE 1—*Erythema Doses and Quality Factors for Xray Exposure*

KV	Filter (mm)	HVL (mm Al)	Erythema Dose (R units)	Quality Factor
75	—	1.0	—	—
100	—	1.5	450	3.1
130	—	2.0	500	2.8
130	4 Al	4.5	700	2.0
130	0.25 Cu; 1 Al	8.0	900	1.6
130	0.5 Cu; 1 Al	11.0	1100	1.3
200	0.75 Cu; 1 Al	15.0	1300	1.1
Harder x rays up to Ra gamma rays			1400	1.0

essary to produce the same biological reaction on the human skin vary markedly with the radiation quality used. The number of R units per erythema decreases with increasing wave lengths." (See Table 1.) The values given in the table are for single dose fractions. Note that the Quality Factor was not defined at the time of this work. Meyer and Glasser observed that for fractionated doses or lower dose rates (less intensity) larger R-unit values would be required to produce the same amount of erythema.

The serious and destructive effects of x rays on human tissue had been recognized for a long time, but there was no well-defined physical quantity to which they could be related. The quantity of x rays needed to produce skin erythema showed great discrepancies when measured with instruments that were calibrated using the various units that had been proposed. Mutscheller (5) related the erythema dose to a quantity defined by the x-ray tube current, exposure time and distance from the tube. He then proposed that the "tolerance dose" be 1/100 of an erythema dose per day. Similar recommendations for the tolerance dose were made by Barklay and Cox (6) who defined a "Unit Skin Dose" (USD) as equal to 8 min of exposure at 9 in from an unfiltered x-ray tube, and 13 h 40 min at the same distance with a 0.5 mm lead filter in place. Their recommended safety limit was 0.00028 USD/day based on the observation that 0.007 USD is "well within the limits of safety". Mutscheller's 1/100 erythema dose can be equated to 0.00033 USD.

In 1927 Behnken (7) proposed a definition of "dose" as the x-ray energy absorbed per unit mass of tissue. The energy absorbed would be determined from the product of the energy fluence and an inter-action coefficient related to the energy absorption.

Fricke and Glasser concentrated on the ionization chamber design problems between 1925 and 1927 (1). They constructed a small chamber with walls that had the same absorption coefficient as air, and could directly measure ionization in air. By 1927 Glasser had perfected the design to eliminate interfering effects of the chamber wall, electrode and position in the field. This allowed the standardization of a measurable quantity that might be related to a biological effect. The unit for this quantity was defined in 1928.

Defining and Re-defining the Roentgen

In the United States the Radiological Society of North America (RSNA) had formed a Standardization Committee to address the radiation dosimetry issue. In 1925 that committee recommended the adoption of the "e unit" as the measure of radiation intensity. The definition was essentially the same as that which was eventually adopted as the definition of the "roentgen" and was based on a previous proposal made by Behnken for what he called the "R" unit (different from Solomon's R unit). The recommendation of the Standardization Committee included the caution that backscattering must be considered when determining skin reaction. A first degree skin reaction was equated to about 1300 e units, but it was noted that this value is dependent upon radiation quality (8).

The roentgen was finally defined and adopted as the international unit for specifying the intensity of x radiation at the 2nd International Congress of Radiology in 1928 (9). However, the physical quantity to which this unit applied was not defined at this time. The application of radiation in the treatment of disease, more than the concern for radiation protection, continued to force further development of the quantities and units pertinent to radiation dosimetry.

In the decade following the definition of the roentgen, L.H. Gray worked on developing an absolute method for measuring radiation intensity using the ionization technique. About the same time as the roentgen was defined, Gray had published the relationship between ionization and energy lost in air, as well as a relationship between the energy lost in air and that lost in a solid (10). He showed that an electron loses the same energy E while traversing a distance Δx in a solid as it would lose in traversing a distance $\rho \Delta x$ in air, where ρ is a proportionality constant independent of particle velocity. From this he derived that

$$E = \rho^2 E'$$

where E is the energy lost by an electron in air and E' is the energy lost by the same electron in a solid. He also went on to show that the number of ions N generated per unit volume is given by

$$N = E'/W$$

where W is the mean energy required to produce an ion pair in the material. It turns out that Bragg had published the same relation in 1912.

By 1936 Gray (11) had detailed the development of the relationship between energy lost E and ionization per unit mass J,

$$E = JW\rho$$

and specified the conditions under which this equality holds. In a subsequent publication (12) he elucidated further the relationship between ionization in a cavity and the energy absorbed in a medium, quoting experimental results from cavities of different sizes and wall materials. He also made some comparisons with calorimetric measurements. Because of this work, Gray argued for a redefinition of the roentgen, in particular to remove the requirement that "the wall effect of the chamber be avoided".

During the same period other investigators were continuing to study the relationship between exposure to radiation and biological effects. Among these, Failla and Henshaw (13) were among the first to standardize on the use of the roentgen as the measure of the amount of radiation for both x rays and radium gamma rays. They investigated the effects of radiation intensity, exposure time and quality on several biological end points and concluded that "there can be no unit of radiation which expresses its biological effectiveness irrespective of quality...".

By the end of 1936 the RSNA Standardization Committee had declared that the roentgen is the fundamental unit of dose. However, it was still not clear whether scattered radiation should or should not be included in the measurement. Both Lauritsen and Taylor expressed concern about the confusion among radiologists in the use of the roentgen as a dose quantity rather than as an intensity unit. They were of the opinion that two units were needed and, in Taylor's mind, that they be related to mass instead of volume (8).

This, then, was the situation in 1937 when the ICRU (14) recommended, and the 5th International Congress of Radiology adopted, a new definition of the roentgen which included gamma radiation, eliminated the requirement for avoiding the wall effect and specified the mass of air rather than a volume of 1 cm^3. This opened up the roentgen for use as a measure of quantity of higher energy photons, and created the possibility of using air cavities for measurement (15).

Beyond the Roentgen and on Toward Dose

Events that took place during the Second World War greatly increased the need for attention to radiation dosimetry for purposes of radiation protection. Even so, the impetus to move beyond the roentgen came primarily from the requirements of radiation therapy. By the late 1940's, with the advent of the betatron, medical radiation therapy had moved into the multi-MeV region. This resulted in the need to evaluate the meaning of local energy transfer and deposition for electrons with long ranges. Also, it was clear that differences had arisen between the physics and medical communities concerning the use of "dose" which led to an attempt to develop some consistent terminology and to separate "field" measurements from dosimetry. Energy absorbed in tissue was recognized as an important quantity and Cantril and Parker (16) had introduced the roentgen equivalent physical (rep), the energy absorbed in tissue exposed to 1 roentgen. At the same time they defined a unit called the rem. Its relation to a physical quantity depended on the radiation and the particular biological effect under consideration. It was meant to specify a quantity of energy absorbed which produced an equivalent effect regardless of the character of the incident radiation.

At a meeting in late 1947 the American College of Radiology (ACR) attempted to take the lead to rationalize a system of dosimetry. G.C. Laurence pointed out that "dose" measured in roentgen is a function of the particle fluence at a point independent of the material, while "energy absorbed" is dependent on the absorbing material. Thus, he argued that, although energy absorbed per gram is important, it should not be called "dose". He felt that a precise conversion should be specified to relate energy absorbed in any material to the dose quantity, and suggested 100 erg/g (8).

During the same meeting L.S. Taylor stated, "I consider it to be the urgent necessity to have a unit which will measure in common terms x rays, γ rays, neutrons, protons and electrons, since for biological purposes, we are compelled to draw comparisons between the effects of these radiations." However, although the rem had been introduced for that purpose, there was some sentiment that, because it was not physical, it was not appropriate at that time to offer it as a unit. Ultimately, no specific recommendations were forthcoming from that meeting (8).

The following year the British Committee for Radiological Units proposed the J unit = 1.58×10^{12} ion pairs/g of air and equivalent to 93 erg/g of energy absorbed in soft tissue. This was essentially the rep of Cantril and Parker. In response, a group of U.S. physicists rejected the J unit because it was not applicable to all types of radiation, specifically neutrons and particles, and it would be subject to variations in stopping power ratios. These physicists, including, Corrigan, Failla, Fano, Laurence, Rossi, Taylor and Wyckoff had come to the conclusion "that the principles of radiation dosimetry must be based upon a fundamental or 'significant' unit". They expressed a conviction that the "primary term of reference for the quantitative characterization of a radiation treatment should be the 'energy dose' ... expressed in standard energy units", realizing that direct measurements of this quantity are not in general practicable (8).

These ideas crystalized over the next few months, and in February 1949 the Committee on Radiological Units, Standards and Protection (CRUSP) of the ACR recommended that a practical unit be specified for a quantity that is defined as the dose of ionizing radiation at a given point in a medium when the energy absorbed is 93 erg/g at the point of interest. However, this was qualified by the accompanying remark that since this quantity, related to an exposure of 1 roentgen, was dependent upon the absorbing medium, perhaps an arbitrary value of 100 erg/g would be preferred.

In 1950 the ICRU, which had not met since before the war, was reconstituted. One of their first recommendations (17) was that dose be expressed as energy absorbed (imparted) per unit mass. However, this did not completely remove the confusion in terminology since it conflicted with their own definition of dose stated in 1937—Dose of x or γ rays is preferably expressed in roentgens (R). The work of the ICRU over the last, almost, 40 years has concentrated on clarification of the dose quantity and its relation to radiation field measurements.

Evolution of Dosimetry—1950 to the Present

The expression of dose as energy absorbed per unit mass did not immediately solve all the problems of dosimetry. There was still an implied, if not expressed, relationship to the roentgen in the concept of the roentgen equivalent. This made the dose quantity dependent upon the value of W/e for air which at that time was uncertain by about 5%, as well as the calculation of energy absorbed in tissue. In 1953 the ICRU (18) dealt with this issue by formalizing the suggestion that Laurence made in 1947 and defined the **rad = 100 erg/g** as the unit for the quantity "absorbed dose". Thus, dose was finally made independent of the roentgen, and consequently, independent of W/e in air.

Further clarifications were made in 1956 (19) with the definitions of the additional quantities "exposure dose" and "RBE dose". This latter quantity followed a recommendation by the National Committee on Radiation Protection (NCRP) in 1954 that RBE be used in the field of radiation protection to provide a mechanism for the addition of absorbed doses of different kinds of radiation. The definitions of these quantities were detailed more precisely in the 1959 report of the ICRU that was issued as NBS Handbook 78 (20). The proportionality between exposure dose R and absorbed dose D was specified under conditions of charged particle equilibrium (CPE) as

$$D = f R,$$

where f is a function of the radiation quality and the absorbing material.

The effects on the absorbed dose of interfaces between materials of different composition, such as soft tissue and bone, were recognized and the report contains information on the absorbed dose per unit exposure dose for various x-ray energies and bone cavity sizes (Fig. 1). Attention was focused on the photoelectrons generated in bone. However, note was taken that at high energy "the pair production process results in greater absorbed dose in compact bone than in soft tissues" (20). The additional dose to adjacent tissues would be between 0 and 15 percent at 10 MV, and the effect could extend to 2 to 3 cm from the interface because of the secondary electron range. This latter point has become more important in recent years with the increasing use of higher energy x rays in radiation therapy.

In the last 25 years, five reports of the ICRU have progressively developed the concepts involved in the absorbed dose and its relationship to the radiation field, which may be external or internal to the body (21–25). During the same period, the International Commission on Radiation Protection (ICRP) and the NCRP have also been active in developing concepts for the application of these quantities to radiation protection from both external and internal sources of radiation, in the measurement of the radiation field quantities, and in the conversion of the measured quantities to the appropriate dose quantities (NCRP (26–28)) (ICRP, (29–32)).

A. External Radiation

Ionization in air cavities has emerged as the preferred measurement method for the dosimetry of external radiation fields. For many years this method was referenced to exposure standards specified in roentgens. The absorbed dose in air, D_{air}, was defined as

$$D_{air} = 0.86_9 X,$$

where x was the exposure in roentgens. The proportionality constant depended, of course on the value of W/e in air. The ICRU (21) defined W/e = 33.7 eV/ion pair (33.7 J/C), which remained the accepted value

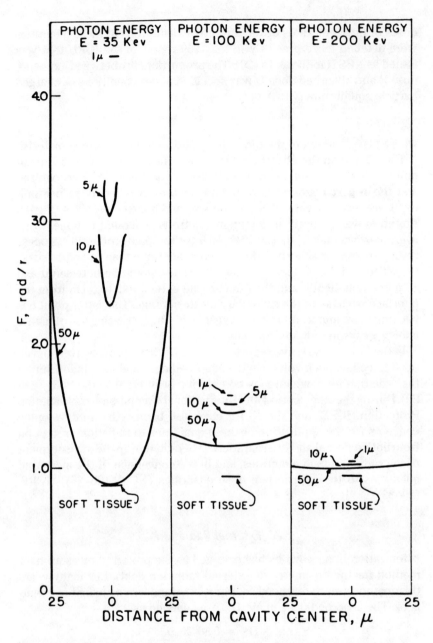

Fig. 1. Dose to soft tissue inside bone cavities per unit exposure for various photon energies. The dose is shown as a function of the distance from the center of the cavity for cavities of 1, 5, 10 and 50 μm diameter. (From Ref. 20).

until 1979 when it was specified as 33.85 J/C for dry air (33). Recently many experimental determinations of W/e were reevaluated using the stopping power information given by ICRU (34) and a new value of W/e = 33.97 ± 0.05 J/C was proposed by Boutillon and Perroche-Roux (35). Absorbed dose to a medium was determined from a measurement of exposure and the appropriate mass stopping power and mass energy absorption coefficient ratios, as well as a correction for the perturbation of the radiation field by the cavity and its wall.

$$D_{med} = D_{air} \, S/\rho)_{w,a} \mu_{en}/\rho)_{m,w} P,$$

where $S/\rho)_{w,a}$ is the stopping power ratio of the cavity wall to air, $\mu_{en}/\rho)_{m,w}$ is the energy absorption coefficient ratio of the medium to the cavity wall, and P is the perturbation correction.

The definitions of absorbed dose and exposure (dose was dropped in 1971) were refined in the 1970's (22–23), and the RBE dose was renamed "dose equivalent" H which was related to the absorbed dose D by the quality factor Q.

$$H = DQ.$$

Q was said to be a function of the collision stopping power of the ionizing particles.

Also, the ICRU (23) introduced the Index Quantities. The absorbed dose index and the dose equivalent index were defined as the maxima of these quantities in a 30 cm diameter sphere, and specified for a point at the center of the sphere. The idea was to have a simple unambiguous specification of the radiation field that was applicable and useful in radiation protection. Some measurement difficulties remained, however, because instruments that are calibrated in a unidirectional field will overestimate the index quantities if they are used in multi-directional radiation fields.

The great advantage of the quantity absorbed dose is that it is independent of the radiation. Thus, the quantity can be applied to measure particulate radiation as well as x and γ rays. With the definition of a biological dose quantity, dose equivalent, the dependence upon the radiation was again manifested because the quality factor is radiation and energy dependent. Energy dependence of Q is minimal for electrons set in motion by x and γ rays, but it varies significantly over the energy range found in typical neutron radiation fields. Consequently, a major problem for radiation protection is the determination of dose equivalent for neutrons.

The NCRP (36) published neutron fluence-to-dose equivalent conversions based upon calculations made by W. Snyder in the mid-1950's.

TABLE 2—*Quality Factors*

Neutron Energy	Q		
	NCRP 25	NCRP 38	ICRU 40
Thermal	3	2	—
0.1 MeV	8	7.5	17.5
0.5 MeV	10	11	23.3
1.0 MeV	10.5	11	20.8
2.5 MeV	8	9	15
5.0 MeV	7	8	10.5
10 MeV	6.5	6.5	7

NCRP Report 38 (28) revised the conversions somewhat, but they have remained unchanged until now. The recent publication of ICRU Report 40 (37) which redefines the quality factor will undoubtedly cause changes in the energy dependence and values of Q that will impact on the evaluation of dose equivalent for neutron fields. The significant conclusion by the ICRU (37) is that the biological effectiveness of radiation correlates more closely with the lineal energy than with the collision stopping power, L∞. In fact, the RBE appears to be proportional to the dose mean of lineal energy. It is a stochastic quantity whose distribution is independent of absorbed dose. Table 2 compares the quality factors based on information in the three reports cited.

New fluence-to-dose equivalent conversion factors have not been published, but the recommendation for an average value for Q to be used for radiation protection purposes has been increased from 10 to 20 by the NCRP (38) and ICRP (39) and to 25 by the ICRU (37).

The introduction of the quantity effective dose equivalent, H_E, by the ICRP (31) allowed partial body radiation exposures to be equated to total body exposures on the basis of risk to the individual. H_E was defined by the equation

$$H_E = \sum_T w_T H_T$$

where H_T is the dose equivalent to a given tissue and w_T is the weighting factor for that tissue representing the proportion of the stochastic risk resulting from tissue T to the total risk when the whole body is irradiated uniformly.

Weighting factors that total 0.7 are specified for certain specific organs. The final factor of 0.3 is to be divided equally among the 5 unspecified organs receiving the highest dose equivalents, not those with the greatest sensitivities. This procedure is somewhat inconsistent with the risk basis used to develop the weighting factors. Gibbs (40) has compared the H_E with a weighted dose equivalent H_W for very non-uniform exposures in diagnostic radiology. The H_W was determined by weighting the dose equivalent for all organs, other than those with specified weighting factors, with 0.3, rather than selecting the 5 organs with the highest dose equivalents. His conclusion is that the H_E probably overestimates the risk of exposure if the exposure is highly non-uniform, *e.g.* confined to a portion of the head.

In recent years the ICRU has focused on defining readily-determinable operational quantities for all types of radiation and has introduced the "ambient", "directional" and "individual" dose equivalents as quantities that can serve for the practical determination of dose equivalent (24, 25). Part of the problem with the index quantities introduced in 1976 is that they are not additive because the maximum dose equivalent can occur at different locations in the body depending on the radiation type, energy, etc. The new quantities are defined in such a way as to overcome that difficulty.

The ambient dose equivalent $H^*(d)$ is defined for a unidirectional radiation field and is usually specified at d = 10 mm. The directional dose equivalent $H'(d)$ is used for specifying superficial exposures and usually d = 0.07 mm. Individual dose equivalents, penetrating $H_p(d)$ and superficial $H_s(d)$ specify the dose equivalent in tissue below a specified point on the body at depth d. For H_p, d is usually 10 mm, while for H_s, d is usually 0.07 mm. The quantities measured by instruments, $H^*(d)$ and $H'(d)$, and personal dosimeters, $H_p(d)$ and $H_s(d)$, can be related to tissue dose equivalent H_T and effective dose equivalent H_E by calculations using appropriate models. Verification of these computations is, however, difficult and in almost all cases the experimental data are lacking.

Relationships among the various radiation quantities are shown for photons in figures 2–4 (25). Figures 2 and 3 demonstrate that the measurement of the radiation field quantity $H^*(10)$ always overestimates H_E except possibly for exposures at photon energies above 10 MeV. Figure 4 illustrates that the placement of the dosimeter to measure $H_p(10)$ is critical to an accurate estimate of H_E. In general, $H_p(10)$ will exceed H_E if the dosimeter is placed appropriately on the body relative to the radiation field. Figures 5 and 6 show the same information for neutron radiation (25).

B. Internal Radiation

Dosimetry of internally-deposited radionuclides has been a concern in radiation protection since the discovery of radon and its potential association with high lung cancer incidence in underground miners, and the observed increased incidence of bone cancer in radium dial painters.

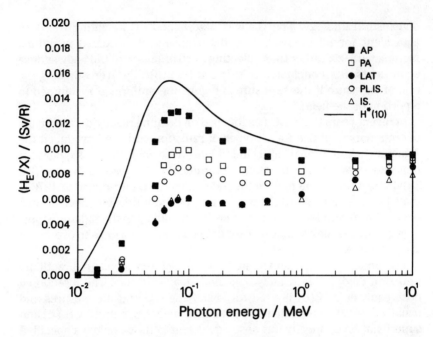

Fig. 2. Effective dose equivalent, H_E, per unit exposure (in free air), X, as a function of photon energy. The solid line gives H^* (10) /X. Five geometries are considered in the calculations: AP, broad parallel beam from anterior to posterior; PA, broad parallel beam from posterior to anterior; LAT, broad parallel beam from the side (lateral); PL.IS, planar isotropic field, perpendicular to body axis; IS, isotropic field. (From Ref. 25).

A method for γ-ray dosimetry was proposed by Marinelli, et al. (41) based on a dose rate constant, a geometrical factor that incorporated the attenuation coefficient and the fraction of activity that disintegrates in a prescribed time. This method was refined and extended for use with β particles by Loevinger, et al. (42). NCRP Report 22 (26) and the ICRP Committee II Report (29) rationalized the approach to this problem. At that time it was recognized that there was a lack of information about the behavior of chemical compounds containing radioactivity once they had been ingested or inhaled. Good human data on exposure to internally-deposited radioactivity was limited to the ingestion of radium, information from a few accident cases and some therapy patients who had been given specific radionuclides in specific compounds. Consequently, it was necessary to assume that the normal stable distribution of an element in various body organs is typical of the distribution that would exist following intake of radionuclides of the same element.

There was little information about the biological variability in humans. The biological retention (half life) and elimination from various organs

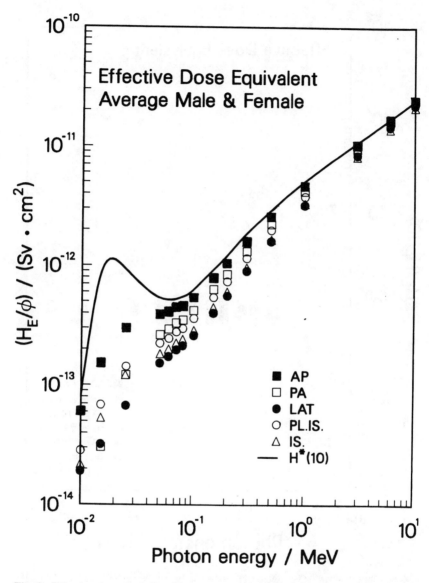

Fig. 3. Effective dose equivalent, H_E, per unit fluence as a function of photon energy. Geometries and symbols are explained in the caption to Fig. 2. The curve for the ambient dose equivalent, $H^*(10)$, is inserted for comparison. (From Ref. 25.)

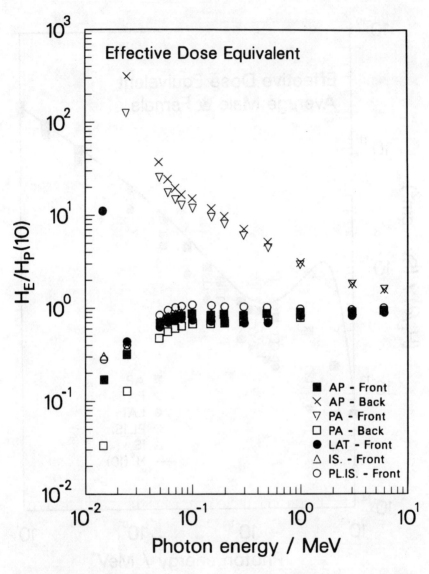

Fig. 4. Ratio of the effective dose equivalent, H_E, to the individual dose equivalent, $H_p(10)$, as a function of photon energy. Two locations for the personal dosimeter are considered: front of the body (Front) and back of the body (Back). $H_p(10)$ is approximated by the dose equivalent at depth 10 mm along the central axis in the ICRU sphere. Geometries and symbols are explained in the caption to Fig. 2. (From Ref. 25.)

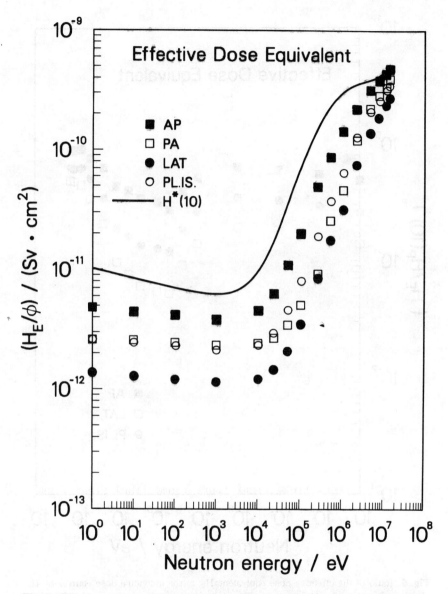

Fig. 5. Effective dose equivalent, H_E, per unit fluence as a function of neutron energy. Geometries and symbols are explained in the caption to Fig. 2. The curve for the ambient dose equivalent, $H^*(10)$, is inserted for comparison. (From Ref. 25.)

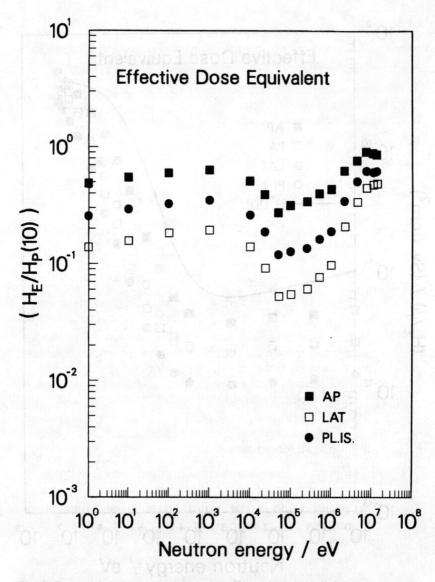

Fig. 6. Ratio of the effective dose equivalent, H_E, to the individual dose equivalent, H_p (10), as a function of neutron energy. The personal dosimeter is considered to be worn on the front of the body. H_p (10) is approximated by the dose equivalent at depth 10 mm along the central axis in the ICRU sphere. Geometries and symbols are explained in the caption to Fig. 2. (From Ref. 25.)

was not known in most cases. This necessitated many assumptions and the use of a specified "standard man" for dosimetry. For each radionuclide analyzed a "critical organ" was designated on the basis of:

1. Accumulation in the organ (predominant factor),
2. Essentialness of the organ,
3. Damage to the entire body as a result of damage to the organ,
4. Radiosensitivity of the organ.

Biological dose was calculated in units of rem = rad x RBE, with

RBE = 1   for β, γ and x radiation and conversion electrons with LET < 3.5 keV/μm.
= 1.7 for low energy β emitters (E < 0.03 MeV).
= 10   for α particles.
= 20   for recoil nuclei.

The ICRP Committee II Report (29) defined an "effective energy deposited per disintegration",

$$\epsilon = \Sigma EF(RBE)n.$$

where E = energy deposited/disintegration;
F = ratio of disintegrations of decay products to disintegrations of parent;
n = a relative damage factor (relative to the effectiveness of radium in the skeleton).
n = 1 if the parent is an isotope of radium, or if the energy emitted originates as x or γ radiation.
n = 5 in all other cases.

The dose D to a given organ was calculated from,

$$D = K\, qf_2\, \epsilon/m,$$

where m = organ mass;
f_2 = fraction of total body burden in the organ;
q = total activity in the body;
ϵ = effective absorbed energy/disintegration in the organ;
K = a constant to adjust the units.

If the organ burden is not constant,

$$qf_2 = P\{1.44\, \tau\, [1-\exp(-0.693t/\tau)]\}$$

where P = the rate of uptake;
τ = the elimination of half-time.

This dosimetry formalism was employed to evaluate dose to individ-

TABLE 3—*Comparison of dose calculated to various organs using ICRP II and MIRD*

Organ	(mSv/10^{14} particles)							
	E γ (MeV)						^{90}Y	
	0.1		0.5		1.0			
	ICRP	MIRD	ICRP	MIRD	ICRP	MIRD	ICRP	MIRD
thyroid	60.8	22.4	384	133	704	240	7200	7520
kidney	9.0	3.8	54.4	20.8	104	36.8	480	480
liver	2.2	1.5	13.3	7.0	25.6	13.0	84.8	83.2

uals from internally-deposited radioactivity as well as to calculate the allowable concentrations of radionuclides in air and water for radiation protection purposes.

About 1968 the MIRD (Medical Internal Radiation Dose) method began to evolve (43, 44). Once again the primary motive was to make better estimates of the radiation dose received by patients in medical applications, rather than for radiation protection of users or the public. A review of this method is found in ICRU Report 32 (45). The physical and biological quantities required to calculate the dose remained the same as in the ICRP method; the formalism changed, and a significant effort was made to accurately calculate the physical parameters such as the energy emitted in the nuclear transitions and the fraction of energy absorbed in the target region.

The biological parameters were grouped into a single term called the *Cumulated Activity* A_h that incorporates the uptake and elimination over time and is analogous to qf_2. The physical terms were eventually grouped into a single *S-Factor* that is the product of the mean energy per nuclear transition Δ_i and the specific absorbed fraction ϕ, and is analogous to ϵ/m. The major differences in the organ doses that would be calculated by the method of ICRP Committee II and MIRD arise from the estimates of the fraction of γ-ray energy absorbed. A comparison of dose to thyroid, kidney and liver for γ-ray energies of 0.1, 0.5 and 1 MeV, and ^{90}Y β particles is given in Table 3.

The development of a system for dosimetry of internal emitters continued through the 1970's with the effort concentrated in three areas: (1) the design of detailed mathematical models to represent the body of a "standard man"; (2) the characterization of the intensities and the energy spectra of the various radiations emitted by radionuclides of interest; and (3) the development of transport methods for calculating the transfer of energy from source to target (46). Ellett, Brownell and their colleagues (47, 48) began work on the problem from the perspective of the medical uses of radionuclides (MIRD). Snyder and his colleagues followed with calculations of absorbed fractions for the MIRD Committee and then used the approach to address the problem from the perspective of radiation protection which culminated in ICRP Publication 30 (32) and its massive volume of supporting data. The final result of these efforts is that the MIRD and the ICRP methodologies have been reconciled and now the ICRP dose calculations give the same values as the MIRD calculations in Table III for the organ dose for γ rays.

The general equation used to obtain dose equivalent rate to a target organ j from activity $A_i(t)$ in source organ i in the ICRP 30 formalism is:

$$\dot{H}_j(t) = \Sigma_i\, S_{ij}\, A_i(t),$$

which is basically the same equation as that used in the MIRD formalism. The difference is that ICRP uses Specific Effective Energy (SEE) (30) which is essentially equivalent to S_{ij} and the MIRD S-factor except that it includes a quality factor. It should be noted that the values of S_{ij} or SEE are dependent upon age and sex. Some work has begun to develop age and sex-dependent values.

In the current ICRP (32) formalism there are no dose modifying factors other than the quality factor Q, and there is no requirement for the non-uniform distribution factor n that appeared in the previous models for bone dosimetry. ICRP no longer recommends that a continuous intake be used to determine limits for internal emitters because the pattern of uptake has little effect on the long-term accumulation of dose equivalent. Time integrals of dose following a single intake should be used instead. However, the intake pattern does influence the short-term dose and may be important to the assessment of specific exposures.

Biological and physiological models are required for estimating uptake and retention and the evaluation of $A_i(t)$. The simple lung model used by the NCRP in 1959 was replaced with a more sophisticated model (49) and modified again (50) before being applied in ICRP Publication 30 (32). The lastest modification changed some of the clearance constants and grouped the tracheo-bronchial region (T-B), pulmonary region (P) and lymph nodes (L) into a single compartment with a mass of 1 Kg, because of uncertainties about the locations of the cells at risk. Particles with AMAD greater than 10 μm were ignored because they are relatively unimportant to the dose, and particles with AMAD less than 0.1 μm were not considered because of the difficulty in predicting their behavior.

The model for the GI tract was that developed by Eve (51). It may result in overestimates of the dose from low energy photons, α particles and β particles because of a failure to properly account for the mucus layer and the depth of proliferating cells.

ICRP Publication (29) used a bone model in which the average dose to the skeleton was calculated and the factor n was used to account for

TABLE 4—*Absorbed Fractions for Dosimetry of Radionuclides in Bone*

Source Organ (Bone)	Target Organ	Type of Radionuclide				
		α emitter uniform in volume	α emitter on bone surfaces	β emitter uniform in volume	β emitter on bone surfaces $E_\beta > 0.2 \text{MeV}^a$	β emitter on bone surfaces $E_\beta < 0.2 \text{MeV}^a$
Trab.	Bone Surf.	0.025	0.25	0.025	0.025	0.25
Cort.	Bone Surf.	0.01	0.25	0.015	0.015	0.25
Trab.	Red Marrow	0.05	0.5	0.35	0.5	0.5
Cort.	Red Marrow	0.0	0.0	0.0	0.0	0.0

^aAverage energy of each β decay group.

non-uniform distribution in the determination of dose equivalent. ICRP Publication (32) adopted a much more complex model for bone dosimetry. The model now specifies two target populations: (1) endosteal cells forming a layer 10 μm thick and covering all the bone surfaces and (2) active red marrow occupying the cavities of trabecular bone. Absorbed fractions were developed for various combinations of emitted particle, source and target, and range from 0 to 0.5 (see Table IV, Johnson, (52)). The actual dose to bone or red marrow is critically dependent upon the distribution of the radiouclide. Differences of an order of magnitude or more are possible depending on whether the activity is in the bone volume or on the bone surface.

Johnson (52) gives a good review of the current ICRP internal dosimetry method and the NCRP (53) has issued a report that evaluates the ICRP dose calculational methods and models that were used for internally-deposited radionuclides as described in ICRP Publication 30 (32). This report concludes that the models and methods are superior to those recommended by NCRP in 1959 (26) but that some deficiencies remain and should form the basis for additional research. With respect to the lung model, there is concern about the lack of consideration of the nasopharangeal (N-P) region and the combining of the rest of the regions into a single mass for an average dose calculation. Information that might be related to specific effects in various regions of the lung is lost. Although the NCRP recommends the use of this lung model for occupational radiation protection, it is appropriate only for prospective planning and not for retrospective evaluation of individual doses when the organ burden is known. More sophisticated lung models are used for the estimation of dose to the bronchial epithelium from deposition of radon decay products (54).

NCRP also has some reservations concerning the bone dosimetry model in that the cells adjacent to the endosteal bone surfaces may be only a part of the target cell population for bone cancer. Osteo-progenitor cells may be located several cell diameters away from the bone surface. Finally, bone marrow stem cells are thought to be distributed non-uniformly within the marrow.

There are large uncertainties remaining in the dosimetry of internally-deposited radioactivity because of the mismatch between the body of a given individual and that of the mathematical phantom. This problem is most acute in the dosimetry of photons and not so severe for α and β particles. The radiological physics data on the emission and transport of radiation are known with great accuracy. However, the knowledge of radioactivity distribution in the body and the kinetics of uptake and elimination are lacking, and this results in additional large uncertainties.

The Present

The field of radiation dosimetry has come a long way since the discovery of x rays and radioactivity at the end of the last century. The international bodies, the ICRU and the ICRP, and in the U.S., the NCRP and the MIRD Committee have developed a rational system for the dosimetry of both external radiation exposures and internal radiation emitters. As the understanding of the mechanisms of energy deposition by ionizing radiation and the biological effects of that radiation increases, the sophistication of the dosimetry also must increase. Greater accuracy is demanded both in medical applications and in radiation protection. This requires continual improvement in instrumentation as well as in calculational models and techniques. In particular, Monte Carlo simulations and analog transport calculations will become increasingly important as one attempts to determine dose in transition zones near interfaces, develop age and sex-dependent models for specific organ doses, evaluate specific absorbed energy at scales approaching the cell, or even the scales of macromolecules. Combined with these demands will be the necessity of making accurate measurements of some "benchmark" models to validate the calculations.

A. Dosimetry of External Radiation Fields

The ICRU has defined the field quantities and has provided information on the relationship between the measured quantities and the dose equivalent to specific organs or effective dose equivalent. These relationships are based on radiation transport calculations in male and female heterogeneous phantoms and the results are stated to have standard errors of less than 5% with few exceptions (25). The calculations are for monoenergetic photons and neutrons and indicate that under almost all measurement conditions the effective dose equivalent will be overestimated. This produces a factor of safety for radiation protection purposes, but the accuracy is poor for photons with energies less than 50 keV and for neutrons with energies less than a few MeV.

This indicates that there remains a need for better measuring devices and techniques particularly for low-energy photons and neutrons, as well as for β particles. Present neutron dosimeters lack sufficient sensitivity, especially with the recently recommended increase in the quality factor (55). Their response depends on their orientation to the neutron field, and they have significant energy dependence. Accurate dosimetry is difficult without knowledge of the neutron energy spectra.

Accurate dosimetry of β radiation may be the most difficult task of all, especially in the presence of photon radiation. The β-radiation fields are usually extremely non-uniform and the energy spectrum at the point of interest is not known.

In summary, the computational technology exists to accurately estimate a dose or dose equivalent quantity from a measurement in an external radiation field. The uncertainties and inaccuracies in dosimetry arise from inadequacies in the measurement devices and techniques.

B. Dosimetry of Internally Deposited Radioactivity

Internal measurements of dose in humans are generally too invasive to be practical. Measurements in body cavities are possible, but the corresponding SEE or S-factors to relate these measurements to dose to a specific organ do not exist. Thus, it is clear that the dosimetry of internally deposited radionuclides will continue to depend on calculations based upon some knowledge of the amount and distribution of a particular radionuclide in the body. The greatest uncertainties arise from: (1) inaccuracies in the measurement of radioactivity and its distribution in the body; (2) the lack of knowledge of the biological kinetics of all but a few radiopharmaceutical compounds; (3) reliance on calculations using a standard 70 kg man to derive the specific absorbed fractions.

Measurement inaccuracies are probably the least significant contributor to dosimetry uncertainties in the application of radio-labeled pharmaceuticals for use in medical diagnosis. However, for radiation protection purposes and for some other emerging medical applications uncertainties in the determination of the amount and distribution of incorporated radionuclides can be serious. This is especially true for ingested or

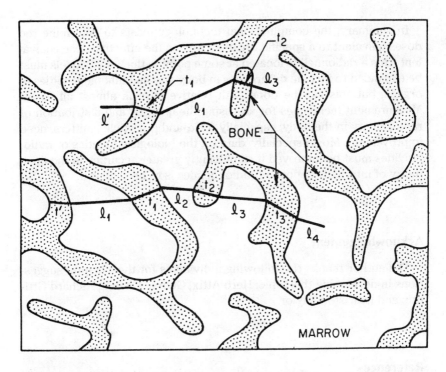

Fig. 7. Model of the bone marrow showing possible electron tracks through the bone mineral and marrow cavities. (From Ref. 56.)

inhaled transuranic elements and other α-particle emitters, and β-particle emitters.

Improving our knowledge of the biological kinetics of radioactive compounds is of highest priority (46). Lack of information in this area is the greatest source of uncertainty in the dosimetry of internally deposited radioactivity in general, and in particular for an individual who may have ingested or inhaled radioactive material.

Calculational models must be, and are being, continually improved to include age and sex-specific phantoms, SEE or S-factors for additional organs, and inhomogenieties in various target organs. The geometry of the bone and bone marrow cavities is especially complex as shown in Figure 7 (56). The transport and energy deposition by electrons in the marrow cavities is crucial for determining dose to the marrow. This can be important in some cases for radiation protection, but it is critical for assessing the impact of the use of radionuclide-labeled antibodies for cancer therapy.

In summary, the computational technology exists to determine the dose equivalent to a specific organ as well as the effective dose equivalent from a radionuclide located at some point in the body. Models must be expanded to include differences in body size and inhomogeneities in organs, but this can be done with relative ease in almost all cases. Measurement techniques for assessing the quantity and distribution of radioactivity in the body, especially for α and β-particle emitters, need improvement. Most especially, data on the biological kinetics of radionuclides must be improved if significantly greater accuracy in the dosimetry of internally deposited radionuclides is to be achieved.

Acknowledgement

The author thanks the following individuals for their helpful suggestions in developing this topic: Herb Attix, George Chabot, Richard Griffith, and Lauriston Taylor.

References

1. Lorenz, E. "The development of dosimetry: An historical sketch," Radiology **14**, 340–345 (1930).
2. Villard, M.P. "Instruments de mesure a lecture directe pour les rayons x," Arch. D'Elec. Med. **16**, 692 (1908).
3. Christen, T. "Radiometry," Arch. Roentgen Ray, **XIX**, 210 (1914).
4. Meyer, W.H. and Glasser, O. "Erythema doses in absolute units," presentation to the Radiological Society of North America, Cleveland (1925).
5. Mutscheller, A. "Physical standards of protection against roentgen ray dangers," Am. A. J. Roentgenol. **13**, 65–70 (1925).
6. Barklay, A.E. and Cox, S. "Radiation risks of the roentgenologist," Am. J. Roentgenol. **19**, 551–561 (1928).
7. Behnken, H. "The German unit of x radiation," Brit. J. Radiol. (Roentgen Society Section). **XXIII,** 72 (1927).
8. Taylor, L.S. *X-ray Measurements and Protection 1913–1964*, National Bureau of Standards Special Publication 625 (National Bureau of Standards, Washington) (1981).
9. ICRU. International Commission on Radiation Units and Measurements. "Recommendations of the International X-Ray Unit Committee," Stockholm Congress issuance, Br. J. Radiol. **1,** 363 (1928).
10. Gray, L.H. "The absorption of penetrating radiation," Proc. Royal Soc. **A122,** 647 (1928).

11. Gray, L.H. "An ionization method for the absolute measurement of gamma-ray energy," Proc. Royal Soc. **A156,** 578 (1936).
12. Gray, L.H. "Radiation dosimetry," Brit. J. Radiol. **10,** 600 and 721 (1937).
13. Failla, G. and Henshaw, P.S. "The relative biological effectiveness of x rays and gamma rays," Radiology **17,** 1, (1931).
14. ICRU. International Commission on Radiation Units and Measurements. "Recommendations of the International Committee for Radiological Units (Chicago, 1937)," Am. J. Roentgenol. Radiat. Ther. **XXXIX,** 295 (1938).
15. Wyckoff, H.O. "From 'quantity of radiation' and 'dose' to 'exposure' and 'absorbed dose'—An historical review," (National Council on Radiation Protection and Measurements, Bethesda, Maryland) (1980).
16. Cantril, S. T. and Parker, H. M. "The tolerance dose." USAEC MDDC-1100 (Technical Information Division, Oak Ridge, Tennessee) (1945).
17. ICRU. International Commission on Radiation Units and Measurements. "Recommendations of the International Commission for Radiological Units (London, 1950)." Am. J. Roentgenol. Radiat. Ther. **LXV,** 99 (1951).
18. ICRU. International Commission on Radiation Units and Measurements. "Recommendations of the International Commission for Radiological Units," Radiology **62,** 106 (1954).
19. ICRU. International Commission on Radiation Units and Measurements. *Report of the International Commission on Radiation Units and Measurements, 1956,* National Bureau of Standards Handbook 62 (U.S. Government Printing Office, Washington) (1957).
20. ICRU. International Commission on Radiation Units and Measurements. *Report of the International Commission on Radiation Units and Measurements, 1959,* National Bureau of Standards Handbook 78 (U.S. Government Printing Office, Washington) (1961).
21. ICRU. International Commission on Radiation Units and Measurements. *Physical Aspects of Irradiation,* ICRU Report No. 10b (International Commission on Radiation Units and Measurements, Bethesda, Maryland) (1964).
22. ICRU. International Commission on Radiation Units and Measurements. *Radiation Quantities and Units,* ICRU Report No. 19 (International Commission on Radiation Units and Measurements, Bethesda, Maryland) (1971).
23. ICRU. International Commission on Radiation Units and Measurements. *Conceptual Basis for the Determination of Dose Equivalent,* ICRU Report No. 25 (International Commission on Radiation Units and Measurements, Bethesda, Maryland) (1976).
24. ICRU. International Commission on Radiation Units and Measurements. *Determination of Dose Equivalents resulting from External Radiation Sources,* ICRU Report No. 39 (International Commission on Radiation Units and Measurements, Bethesda, Maryland) (1985).
25. ICRU. International Commission on Radiation Units and Measurements. *Determination of Dose Equivalents from External Radiation Sources—Part 2,* ICRU Report No. 43 (International Commission on Radiation Units and Measurements, Bethesda, Maryland) (1988).

26. NCRP. National Council on Radiation Protection and Measurements. *Maximum Permissible Body Burdens and Maximum Permissible Concentrations of Radionuclides in Air and in Water for Occupational Exposure*, NCRP Report No. 22 (National Council on Radiation Protection and Measurements, Bethesda, Maryland) (1959).
27. NCRP. National Council on Radiation Protection and Measurements. *Stopping Powers for Use with Cavity Chambers*, NCRP Report No. 27 (National Council on Radiation Protection and Measurements, Bethesda, Maryland) (1961).
28. NCRP. National Council on Radiation Protection and Measurements. *Protection Against Neutron Radiation*, NCRP Report No. 38 (National Council on Radiation Protection and Measurements, Bethesda, Maryland) (1971).
29. ICRP. International Commission on Radiological Protection. *Report of Committee II on Permissible Dose for Internal Radiation*, Publication 2 (Pergamon Press, Oxford) (1959).
30. ICRP. International Commission on Radiological Protection. *Report of the Task Group on Reference Man*, Publication 23 (Pergamon Press, Oxford) (1975).
31. ICRP. International Commission on Radiological Protection. *Recommendations of the ICRP*, Publication 26 (Pergamon Press, Oxford) (1977).
32. ICRP. International Commission on Radiological Protection. *Limits for Intakes of Radionuclides by Workers*, Publication 30 (Pergamon Press, Oxford) (1979).
33. ICRU. International Commission on Radiation Units and Measurements. *Average Energy Required to Produce an Ion Pair*, ICRU Report No. 31 (International Commission on Radiation Units and Measurements, Bethesda, Maryland) (1979).
34. ICRU. International Commission on Radiation Units and Measurements. *Stopping Powers for Electrons and Positrons*, ICRU Report No. 37 (International Commission on Radiation Units and Measurements, Bethesda, Maryland) (1984).
35. Boutillon, M. and Perroche-Roux, A.M. "Re-evaluation of the W value for electrons in dry air," Phys. Med. Biol. **32,** 213-219 (1987).
36. NCRP. National Council on Radiation Protection and Measurements. *Measurement of Absorbed Dose of Neutrons and Mixtures of Neutrons and Gamma Rays*, NCRP Report No. 25 (National Council on Radiation Protection and Measurements, Bethesda, Maryland) (1961).
37. ICRU. International Commission on Radiation Units and Measurements. *The Quality Factor in Radiation Protection*, ICRU Report No. 40 (International Commission on Radiation Units and Measurements, Bethesda, Maryland) (1986).
38. NCRP. National Council on Radiation Protection and Measurements. *Recommendations on Limits for Exposure to Ionizing Radiation*, NCRP Report No. 91 (National Council on Radiation Protection and Measurements, Bethesda, Maryland) (1987).

39. ICRP. International Commission on Radiological Protection. *Statement from the Paris Meeting of the ICRP*, Annals of the ICRP, **15**, No. 3, i-ii (1985).
40. Gibbs, S.J. "Influence of organs in the ICRP's remainder on effective dose equivalent computed for diagnostic radiation exposures," Health Phys. **56**, 515–520 (1989).
41. Marinelli, L.D., Quimby, E.H. and Hine, G.J. "Dosage determination with radioactive isotopes. II. Practical consideration in therapy and protection," Am. J. Roentgenol., Radium Ther. **59**, 260 (1948).
42. Loevinger, R., Holt, J.G. and Hine, G.J. "Internally administered radioisotopes," page 801 in *Radiation Dosimetry*, Hine, G.J. and Brownell, G.L., Eds. (Academic Press, New York) (1956).
43. Loevinger, R., and Berman, M. "A schema for absorbed-dose calculations for biologically-distributed radionuclides," MIRD Pamphlet No. 1, J. Nucl. Med. **9**, Suppl. No. 1, 7–14 (1968).
44. Loevinger, R. and Berman, M. "A formalism for calculation of absorbed dose from radionuclides," Phys. Med. Biol. **13**, 205 (1968).
45. ICRU. International Commission on Radiation Units and Measurements. *Methods of Assessment of Absorbed Dose in Clinical Use of Radionuclides*, ICRU Report No. 32 (International Commission on Radiation Units and Measurements, Bethesda, Maryland) (1979).
46. NCRP. National Council on Radiation Protection and Measurements. *The Experimental Basis for Absorbed Dose Calculations in Medical Uses of Radionuclides*, NCRP Report No. 83 (National Council on Radiation Protection and Measurements, Bethesda, Maryland) (1985).
47. Brownell, G.L., Ellett, W.H. and Reddy, A.R. "Absorbed fractions for photon dosimetry," MIRD Pamphlet No. 3, J. Nucl. Med. Suppl. No. 1 (1968).
48. Ellett, W.H. and Humes, R.M. "Absorbed fractions for small volumes containing photon-emitting radioactivity," MIRD Pamphlet No. 8, J. Nucl. Med. Suppl. No. 5 (1971).
49. ICRP. International Commission on Radiological Protection. "Report of Committee II, Task Group on Lung Dynamics, Deposition and retention models for internal dosimetry of the human respiratory tract," Health Phys. **12**, 173 (1966).
50. ICRP. International Commission on Radiological Protection. *The Metabolism of Compounds of Plutonium and Other Actinides*, Publication 19 (Pergamon Press, Oxford) (1972).
51. Eve, I.S. "A review of the physiology of the gastrointestinal tract in relation to radiation dose from radioactive material," Health Phys. **12**, 131 (1966).
52. Johnson, J.R. "Internal dosimetry for radiation protection," in *The Dosimetry of Ionizing Radiation*, Kase, K.R., Bjarngard, B.E. and Attix, F.H., Eds. (Academic Press, New York) (1985).
53. NCRP. National Council on Radiation Protection and Measurements. *General Concepts for the Dosimetry of Internally Deposited Radionuclides*, NCRP Report No. 84 (National Council on Radiation Protection and Measurements, Bethesda, Maryland) (1985).

54. NCRP. National Council on Radiation Protection and Measurements. *Evaluation of Occupational and Environmental Exposures to Radon and Radon Daughters in the United States,* NCRP Report No. 78 (National Council on Radiation Protection and Measurements, Bethesda, Maryland) (1984).
55. Griffith, R.V. "The next decade in external dosimetry," Health Phys. **55**, 177–189 (1988).
56. Eckerman, K.F. *Aspects of the Dosimetry of Radionuclides within the Skeleton with Particular Emphasis on the Active Marrow,* 4th International Radiopharmaceutical Dosimetry Symposium, CONF-851113, 514–534 (Oak Ridge, Tennessee) (1985).

Discussion

JACK SHAPIRO (Harvard): First of all, I'd like to second Charlie Meinhold's recommendation that perhaps it's time to go back to field quantities and measurement. Particularly with regard to the dose equivalent because when I did my thesis on radon back in 1955, just as I was finishing my thesis, they changed the quality factor for alpha particles from 20 to 10. So I had to go through the whole manuscript and change everything from 20 to 10. When I wrote my book on radiation protection in 1971, just as I had completed the manuscript, they changed the alpha particle quality factor from 10 back to 20. I had to go through the whole manuscript and change everything back from 10 to 20 in the calculations. For the last few years I have been measuring environmental neutron levels using a quality factor of 10 and everybody has been happy because the levels have been around 8 millirem per year. And I've told them that as long as we're below 10 we're fine. Now they've changed the quality factor from 10 to 20 and I'm above 10 millirem per year and I'm not quite sure how I'm going to handle that. So perhaps we can do a little better from that point of view. And finally, God help us if we have to measure the effective dose equivalent. I don't have any problem with effective dose equivalent for radiation protection but if we use effective dose equivalent for measurements we're going to get a lot of useless scientific data which we'll have to change from year to year, even though it's useless, because the numbers will be changing.

RON KATHREN: Jack's a hard act to follow. But I'd like to point out the difficulties, Ken after your very fine talk, of using these quantities and units based on quality factor, etc. for radioepidemiological studies. Many times the results of one study are not comparable with the results of another. And yet it is from these studies that the basic risk estimates are often derived. So I issue that caution which has impact, I think the next speaker will show, in the regulatory process.

Scientific Session

Regulation of Ionizing and Nonionizing Radiation and Implications for the Future

Dade Moeller
Chairman

Development and Current Trends Regarding Regulation of Radiation Protection

L. Manning Muntzing
Doub, Muntzing and Glasgow, Chartered

Introduction

A recent issue of *Time* magazine featured on its cover the question, "Is anything safe?" The magazine was responding to recent concerns about risks regarding Chilean grapes and sprayed apples. We are constantly bombarded today by information designed to show that something we eat, drink or breathe, or some activity in which we are engaged, is or can be potentially harmful to our health.

The problem with so many of these highly publicized concerns is the fact that an individual case is often extrapolated to the entire population or an isolated incident is made to represent a global impact. It should not be long before someone will be able to establish that the scientists who project, the government officials who decide, and the media which report potential risks are all injurious to our health because they are scaring us without justification.

We readily accept many high risks, if these involve a familiar source, such as driving a car over 55 mph, or if they provide some personal benefits, such as owning a gun. We feel we can control these risks because we are in charge; it is we who accept or reject them. Consequently, in these areas we generally have demanded less governmental interference. We are not tolerant of risks, however, when we are not in control and when the benefits are society's and not ours as individuals, especially when the risks are unfamiliar or not easily recognizable and forced upon us, such as with many of the radiological risks.

Is there something about the risk of radiation exposure that is so different from other risks? Because of its origins and the types of risks it represents, the public perceives that there is a difference and this is translated into laws and regulations. This has certainly been true for at least the past twenty-five years. As a result the field of radiation has been widely researched and analyzed. Whether, however, it will ever be considered and compared in the same way with other risks by the public is doubtful without a change in radiation literacy.

It is simple enough to say that what this country needs, and what, in fact, the world should have is objective scientific information, reasonably interpreted, leading to acceptable levels of health and safety which are reported to the public accurately. However, this state of utopia is unlikely to be achieved. When we look at radiation issues and consider occupational and public exposures, it is hard to say that the community of scientists and decision makers have done any worse than in other areas where people analyze and make decisions about a variety of risks. Even so, can we do better to improve the regulation of radiation?

As we examine the lessons of the past twenty-five years and look at emerging trends, there are several concepts that, if used, could be in the best interest of protecting public health and safety with regard to radiation.

> First. The United States government should speak with a consistent and coordinated voice on radiation policies and regulations. This will require a change in the way laws are enacted and implemented.
>
> Second. The United States government and the U.S. scientific community, with the U.S. having the largest nuclear program in the world, should reestablish its leadership in international radiological activities and decisions. This will mean an enhanced U.S. participation in international groups addressing radiation issues and in U.S. policy toward adoption of internationally developed standards.
>
> Third. The relationship between the scientific and political communities should be structured so that interpretation of scientific data is not tempered by political factors and that political decisions are based on scientific data. In the process, scientists need to understand the needs of the political community for certainty and the political community must understand comparability of one risk to another and benefits and costs attributable to risks. This would emphasize scientifically based standards and political decisions using acceptable levels of health and safety as their standard.
>
> Fourth. The radiation literacy of the public should be improved so that sufficient resources are allocated to important risks and that resources

are not wasted on the negligible risks. This could lead to the use of concepts such as below regulatory concern, as well as the effective management of radiation protection.

Fifth. The American love of due process and faith in the judicial system resulting in a litigious society should not be used as a substitute for technological consensus upon which the public interest is based.

With these concepts in mind, there follows a look at the past, a review of where we appear to be headed and, finally some thoughts about where we should proceed in the future.

Historical Perspective

An examination of some key decisions during the past several decades shows events that lead to the suggestions I am making. For instance, during this period the Federal Radiation Council was abolished and its authority transferred to the Environmental Protection Agency. In addition, the Congress has enacted various laws with differing standards or rationale including the Atomic Energy Act, the Clean Air Act, the Clean Water Act, and the Low Level Radioactive Waste Policy Act. The fact is that these actions have resulted in multiple legislative standards, differing concepts of interagency coordination and intermittent ability to have the highest levels of government focus on radiation policies.

Another important development was the publication by the Atomic Energy Commission in 1957 of standards that provided what was considered then to be "a very substantial margin of safety for exposed individuals." In promulgating these standards, the AEC emphasized "that the standards are subject to change with the development of new knowledge, with any significant increase in the average exposure of the whole population to radiation, and with further experience in the administration of the Commission's regulatory program."

Another regulatory milestone occurred in 1970 when the AEC proposed a rule on "Control of Releases of Radioactivity to the Environment" to assure that reasonable efforts were made by all AEC licensees to keep radiation doses and releases of radioactive materials from nuclear reactors to the environment "as low as practicable." The Nuclear Regulatory Commission issued a final regulation in 1975 that established a quantitative standard based not solely on an acceptable level of health and safety but rather on a cost benefit analysis taking into account the state of technology.

A third regulatory milestone occurred in 1974, when the AEC promulgated a new Part 51 entitled, "Licensing and Regulatory Policy and

Procedures for Environmental Protection." This was intended to implement the revised Guidelines of the Council on Environmental Quality pertaining to the preparation of environmental impact statements pursuant to the National Environmental Policy Act. Ten years later, the NRC revised Part 51 which was especially noteworthy with respect to NRC's internal debates concerning the appropriate use of standards for categorical exclusions from environmental reports.

These and other regulations have involved difficult and sometimes inconsistent regulatory decisions about standards and assessment of radiation risks and public exposures. For instance, with respect to the NRC rule concerning disposal of high level radioactive wastes NRC recognized that although EPA, under the ambient environmental standards setting authority, has the responsibility to prepare a standard that will set limits for releases of radioactivity to the general environment from disposal facilities, EPA had no such standard at that time. In the absence of such a standard, NRC examined a range of limits which bound that expected for the EPA standard and attempted to select a proposed performance objective that establishes a release limit for the site boundary in keeping with those parameters.

Changing terminology in the course of radiation standards evolution has been a source of confusion in the scientific community and a barrier to public understanding. Some changes are necessary for scientific reasons; other are perhaps arbitrary and unnecessary. Is ALARA the same as optimized protection? It took several years to sort that out.

The lessons to be learned from these activities are that it is difficult for the scientific and political communities to achieve a harmonious relationship in the radiation field; it is difficult for the government to act consistently and to coordinate decisions since federal agencies themselves do not have a harmonious relationship, the regulatory philosophy being used may not be the best for the public interest; and radiation literacy development is insufficient.

Current Issues

A. Nuclear Regulatory Commission

1. Revision of Part 20

There are two noteworthy developments at the Nuclear Regulatory Commission concerning radiation protection standards: 1) NRC's publication in 1986 of a proposed major revision of its regulations in Part 20; and 2) the proposed issuance of a below regulatory concern (BRC) policy.

The Part 20 revision would be based to a great degree on ICRP recommendations. However, the ICRP recommendations were issued in 1977. Other countries adopted these recommendations many years ago. This inability to act in a timely manner affects the United States' leadership throughout the world in radiation standards.

One of the problems with the Part 20 revision has been that instead of proceeding smoothly to promulgation of a final rule, over 800 comment letters in response to the proposed revised rule indicate that there is disagreement about the justification for the rule; radiation protection principles; the ALARA concept; acceptability of risks; quantification of risks from occupational exposures and from exposures of individuals in the general population; risks for minors and pregnant women; planned special exposures; overexposures; standards for emergency and accident conditions; risks for transient and moonlighting workers; and setting of *de minimis* levels and BRC levels; and individual and collective dose evaluations.

Complexity and cost are certainly not absent from a revised Part 20. The NRC staff envisages that its proposal would have four major impacts: 1) the new rule would be more complex, and licensees would have to spend time and resources in understanding it; 2) the new concepts and terms would require licensees to revise their procedures and their training; 3) licensees would be required to provide annual dose reports to all workers; and 4) fuel fabrication facilities could have significantly increased costs in order to control doses from certain long-lived radionuclides.

Even before publication of the proposed new Part 20, the NRC staff began a more extensive involvement in a federal interagency committee, coordinated by EPA, that is engaged in developing guidance for federal agencies on radiation protection standards for the public. The problem has been that these efforts have revealed inconsistencies in existing radiation protection standards (which will take considerable effort to resolve) and problems in varying attitudes about such matters as distinctions between dose considered to be below regulatory concern and those considered to be *de minimis*. During this interagency review, NRC decided to separate the *de minimis* and BRC issues from its revision of Part 20 and to resolve them in a separate policy statement.

2. Policy on Below Regulatory Concern

On December 12, 1988, NRC published an advance notice of a Policy Statement on Exemptions from Regulatory Control saying that it intends to issue the general policy statement in order to establish a basis upon which it can develop regulations and make practice-by-practice exemption decisions. In proposing 10 mrem as the level of annual individual dose practice-by-practice, it emphasized that its policy does "not assert an absence or threshold of risk but rather establishes a baseline where further government regulation to reduce risks is unwarranted." Thus, its present "practice-specific" exemptions would be encompassed within a broader NRC policy which defines levels of radiation risk below which specified practices would not require NRC regulation.

NRC recognizes that the BRC matter and the numerical criteria ultimately selected are significant and contentious issues internationally and that, therefore, "some degree of consistency internationally is desirable, since exemption decisions can affect populations outside each country's borders."

The key to evaluating NRC's proposal is understanding the Commission's three central concerns with respect to whether or not to grant exemptions from regulatory control. These concerns focus on: 1) the extent to which exposures resulting from any practice exempted from regulatory control should be justified; 2) whether to employ a collective dose as a measure of impact in addition to an individual dose criterion; and 3) whether or not individuals may experience radiation exposure approaching the limiting BRC values through the cumulative effects of more than one BRC practice, even though the exposures from each practice are only small fractions of the umbrella limit.

On January 12, 1989, the NRC staff hosted a public meeting to discuss the proposed BRC policy statement. The opinions seemed to be divided evenly between representatives of the industry who generally supported the need for a policy and those of groups who opposed it. Importantly, although NRC stated that it was aware that EPA's current regulatory criteria are more restrictive than exemptions NRC would grant under the proposed policy, EPA pointed out its own authority regarding radioactive releases to the environment and stated that it did not believe that, as proposed, the policy would protect adequately the public health and safety or the quality of the environment. Moreover, EPA was concerned about the appropriateness of specific elements of NRC's proposed policy, stating that the 10 mrem individual dose criterion is too high and that collective dose or some other measure of societal impact must be included. In addition, it questioned whether NRC should have acted unilaterally, at this time, in view of previous and evolving EPA and NRC policies. Finally, it suggested that efforts be directed case-by-case rather than at generic evaluations and proposed that a more coordinated approach should be developed in which NRC and EPA would work together through a task force or interagency working group to formulate a joint policy.

B. Environmental Protection Agency

From a radiological standpoint, aside from legislative initiatives to renew the Clean Air Act, the most important current regulatory item at EPA is the proposed rule it published on March 7, 1989, on radionuclide National Emission Standards for Hazardous Air Pollutants using the principles of the D.C. Circuit Court's decision in *NRDC v. EPA*, 824 F.2d 1146 (D.C. Cir. 1987) (referred to as the *Vinyl Chloride* decision). EPA is to issue its regulatory decision for all radionuclide source categories by August 31, 1989.

In the *Vinyl Chloride* decision, the Court set out a two-step decision process for EPA to follow in establishing standards. First, it has to determine a "safe" or "acceptable" health risk level. Second, it must set the standard at the level—which may be lower but not higher than the safe or acceptable level—which protects the public health with an ample margin of safety. The rule has been and will continue to be highly contentious, and it is doubtful that the litigation on it will cease.

C. Department of Energy

The National Laboratories of the Department of Energy have been an extraordinary source of scientific expertise through the years and a bastion for scientific development in the field of radiation protection. On the other hand, DOE's production and waste management side has had its problems so that the overall credibility of the Department on radiological matters is a mixed one. If there is a federal agency that could and should lead the effort for public education concerning radiation, it would be the Department of Energy. Unfortunately, when the Atomic Energy Commission was abolished this function was lost in the reorganization. It is a trend without a current solution that needs to be addressed. Education of the public should go hand in hand with the development of technology.

D. Committee on Interagency Radiation Research and Policy Coordination (CIRRPC)

When Congress enacts laws that have different standards, as they have done in the radiation field, or when clear organizational structures do not exist, institutions arise to bridge the gap. A useful development has been the establishment of the Committee on Interagency Radiation Research and Policy Coordination (CIRRPC) which provides a neutral ground for federal agencies to exchange views and coordinate their activities. It is an important function limited by the fact that the Committee is a coordinating body without authority to direct consistent and uniform policies throughout the federal government.

Trends and Future Developments

Having looked at the historical development and current issues, I would like to return to the five concepts introduced at the beginning of this presentation.

With respect to the issue of consistency and uniformity within the United States government, there are efforts to achieve that objective including EPA's interagency committees and the activities pursued by CIRRPC. One of the difficulties is that the Congress continues to enact legislation with differing standards creating implementation problems for those in the government responsible for administering these laws. If nothing occurs with regard to this approach, it will be difficult to overcome inherent problems, even with the best of intentions. In my opinion we should reestablish something comparable to the Federal Radiation Council within the Office of Science and Technology Policy with both sufficient stature and power to undertake appropriate activities. We either must recognize that radiation is a type of special hazard which requires special attention and, therefore, special governmental responses or that it is not so important and can be left very much as it is. Until there is higher radiation literacy in this country, the better approach is to return to the previous concept of the Federal Radiation Council.

On the international scene since Chernobyl we have seen an increased effort of nations, including Eastern European countries, to cooperate on nuclear issues. Some people have described this as a new era of international nuclear exchanges and cooperation. In the field of radiation protection the United States must participate aggressively in international activities and use the developments that emerge, or its leadership role will be stifled. A Federal Radiation Council would help to revive the United States leadership if it had a prime mission of encouraging international participation, such as through the NCRP or with federal/state interagency consensus, and of using in federal regulations internationally developed radiation protection standards.

The scientific and political communities have struggled with radiation protection through the years. There are several basic concepts in this relationship that are important to restate so that they are not forgotten. The first concept is that scientific data and analysis must be objective and where there are uncertainties, those uncertainties must be stated clearly. It is for the scientific community to outline what is known and to analyze the risks and describe the benefits and costs associated with them. It is not for the scientific community to decide what society needs. Its role is to provide data that is accurate and understandable. It is for the political arena, on behalf of the public it represents, to take that data and to decide what should be an acceptable standard of safety and health. The politician should not attempt to be a scientist and take the view that whatever is possible should be done. Instead the political judgment should be used to determine what is an acceptable level of health and safety based on reliable scientific data. Once that level has been determined then that should be the end of the requirements for protection against trivial risks, or in other words below regulatory concern.

Public education to raise the level of radiation literacy has not been widely undertaken by the government since the Atomic Energy Commission was abolished. This is a function that a Federal Radiation Council could provide with the proper mission and resources so that there is enhanced radiation literacy in this country with the public understanding benefits, costs, and comparisons. In addition, it is a role that organizations such as the NCRP, the Health Physics Society, the American Nuclear Society, and Federal and State Radiological Health agencies should undertake. The public respects scientific opinion and when the scientific community speaks, hopefully with one voice, it is well received. The practice of excellence in governmental and commercial radiation protection programs is also essential for public acceptance. Thus, support for national and international radiation literacy programs should come from the private as well as the governmental sectors.

Finally, this nation operates under a constitution with laws and procedures and has a judicial system by which anarchy is avoided and disputes are resolved using concepts of due process so that people who have an interest can be heard and their views considered. The result, however, is that the court system tends to be the final arbitrator of many issues. To their credit the courts have established the doctrine that they will give due deference to the expertise of a regulatory agency both as to its technical decisions and its procedural processes. It should be anticipated that radiological issues will be decided in the courts. Therefore, the thresholds for attacking radiological decisions, technologically justified, should be limited and the Congress should so indicate in its legislation. The best public policy can be built with scientific and technical consensus and not through adversarial procedural processes either within the agencies or within the courts. Where feelings are strongly held, this is a difficult concept to achieve, but the enhancement of both domestic and international scientific consensus building forums that include diverse viewpoints can help to bring about this result.

Conclusion

The people of the United States and throughout the world have been well protected through the years by the scientific effort and the resulting political decisions on radiological protection. It has cost a great deal and in fact a criticism could be that it has cost too much to regulate the industry to an unnecessary degree.

The key issue I raise is whether we can do better and I believe we can. The radiological community has done a good job but it should not be complacent. It is time to better coordinate activities; it is time to regain U.S. leadership in the world; it is time for the scientific and political communities to pull together rationally; it is time to raise the level of radiation literacy; and it is time to emphasize not litigation but scientific consensus.

Radiation Protection Policy: Implications for the Future

Warren K. Sinclair
National Council on Radiation Protection and Measurements
Bethesda, Maryland

This audience has been treated in the last day and a half to a remarkable survey of progress in radiation protection, from its early beginnings soon after the discovery of radiation in 1895, to the present time. A few glimpses of the future have also been provided in the earlier papers in this Proceedings.

In this paper some of the subjects that have been discussed in this meeting will be recapped briefly, with special regard to the most recent results because where we have been and what we have learned has much to do with where we can go in the future. Furthermore in the process of this recap, some of the problems that the NCRP (and others) must address in the future will be indicated.

Scientific Basis of Radiation Effects

From the presentations on the scientific basis of radiation protection; it appears that risk estimates for cancer (Table 1) are about three times higher than they were a decade ago, possibly more, depending on the age of the individuals considered and the value assigned to the dose reduction factor (1, 2). While this is not outside the range of uncertainties risk estimates were known to be subject to in the 1975–1980 period (3, 4), it is a larger change than would have been anticipated at that time. Furthermore, the distribution of risk among organs is somewhat differ-

TABLE 1—*Lifetime risk of radiation induced cancer*

	Lifetime Risk of Cancer, %/Sv	
	High dose, high dose rate	Low dose, low dose rate
UNSCEAR, 1977	2.5	1.0
UNSCEAR, 1988	4–11	(3–5)[+]
	(DREF*, 2–10)	

*DREF = Dose Rate Effectiveness Factor (10)
[+] author estimate

ent than before, as shown in Table 2, comparing UNSCEAR in 1977 with the multiplicative values given in UNSCEAR 1988. Notable differences occur in the ratio of the organ risk to the total, in organs like the breast and the stomach. This could be somewhat misleading however, since the 1977 report gave results based on all human sources of information whereas the 1988 report based its estimates on the Japanese population only. These numbers need some adjustment for other populations such as the U.S. population.

On the other hand, appraisals of well known radiation-induced genetic effects, such as autosomal dominant and recessive changes and chromosomal changes have not changed sensibly recently (2) and may actually be less than the relatively high values ICRP used a decade ago (5). Humans are not more sensitive than the mouse on which the estimates are based, indeed they may be less sensitive (6, 7). One problem here however, is that the complex multifactorial diseases are not accounted for in these estimates and their spontaneous incidence in human beings is now known to be much greater than was believed in 1977. Unfortunately their radiation inducibility is not known and can only be very approximately estimated. This may cause some difficulties for us in the future.

TABLE 2—*Fatal cancer percent per Sv by tissue, high dose, high dose rate*

	UNSCEAR 1977	UNSCEAR 1988 (Multip.)
Bone marrow	0.50	0.97
Bladder	0.05	0.39
Breast	0.50	0.60
Colon	0.15	0.79
Lung	0.50	1.5
Multiple myeloma	—	0.22
Ovary	—	0.31
Oesophagus	0.5	0.34
Stomach	0.15	1.3
Remainder	0.6	1.1
	2.5	7.1

A new facet of the problem of mental retardation resulting from exposure of the fetus has arisen. The risk of mental retardation is now recognized to be most severe during the first 8 to 15 weeks of gestation (about 40% per Sv), less severe in the following period, 16 to 25 weeks, and essentially zero at other times (8). This poses new problems for radiation protection.

In the more fundamental radiobiological area extrapolation from animals to man can be quite satisfactory, even numerically, if done carefully on a relative risk basis. However, it is apparent that allowances for the difference between low dose and dose rate effects and high dose and dose rate effects, critical to risk estimation, are not easy to make (9). Animal experiments comparing dose rate effects for tumor induction or life shortening, the most relevant endpoints, suggest factors between 2 and 10 for low-LET radiations (10). Comparable evidence on dose rate effects is lacking in human beings. One survey (11) recommended, for humans, the use of a factor of "up to 5." UNSCEAR 1988 recommended a factor of from 2 to 10, without however surveying any new data. Much more work, both laboratory and human, is needed in this important area.

In recent times neutrons have been recognized to be more effective at low doses relative to x and γ rays than was originally thought (12, 13). Neutrons now appear to have joined α particles in having a preferred Q value of about 20 (13, 14, 15). However, there are some problems. The baseline reference radiation, whether it is defined as x or γ for example, influences these values by a factor of 2 or 3 but protection organizations have always set all low-LET radiations equal to 1 (*e.g.*, 5, 15). Furthermore, there is increasing recognition that when the entire human body is the system to be protected, dosimetry and the change in Q with depth in the body is also a factor. Thus the application of Q values directly will tend to overestimate the effects of radiations like neutrons in the human body. Modifications in the system could be made although at the price of additional complexity. Some consideration is now being given to this (16).

Perception of Risk

An interesting discussion took place on perception of risk (17). One important outcome is that professionals tend to see the risk of radiation as less than it really is, especially those professionals who are closely associated with radiation work. The public on the other hand, sees it as more hazardous than it really is, mainly because of their fear of the unknowns associated with radiation exposure. So far our efforts to

TABLE 3—*Public exposure in the U.S.: Average annual exposure*

	Natural mSv	Fallout mSv	Medical mSv
1950–60	0.8–1.7*	~ 0.1	0.8–2.8
1975–80	~ 1.0* radon additional	fallout reduced	0.9–1.0
1985–90	3.0 includes radon	neg.	~ 0.5

*Cosmic + terrestrial + internal components

balance the public view have not been very successful but perhaps the professional view needs some balancing as well. The important area of perception, to which NCRP devoted an entire meeting a decade ago (18) deserves much more study.

Experience with Radiation Exposure

Occupational exposure. — The trend in average levels of exposure to individual workers has been steadily downward, while the number of people involved has been increasing. If the current trend continues, about two million people will be involved in radiation work in 1990 and the average exposure of those exposed will be ~ 2 mSv (19, 20, 21).

Radiation protection has not been uniformly effective in all segments of occupational exposure. For example in the nuclear power industry (22) the average dose to a worker is several times the average dose to all other radiation workers, and U.S. experience is notably poorer than that of most other countries. Suggestions have been offered for improving the application of ALARA principles and for reducing exposures overall (22).

In medical radiology on the other hand, occupational exposures continue to improve steadily, possibly along with developments in imaging receptor technology. Special radiological procedures, such as those in cardiology, constitute some of the more important areas of operator exposure. The future may see more persistent application of ALARA procedures in medical applications also (23).

Public exposure. — In the 50's and 60's fallout, peaking at about 0.1 mSv/y, was a significant addition to the natural exposure. The latter was estimated to be about 1 mSv/y from cosmic, terrestrial and internal sources. By 1975 fallout had decreased and a component of radon exposure to the lungs was recognized (24) as shown in Table 3. In the late 1980's fallout has become negligible but indoor radon now dominates

SOURCES OF RADIATION EXPOSURE TO U.S. POPULATION

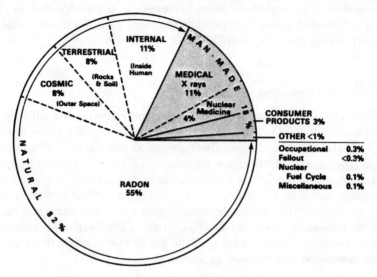

Total Exposure 3.6 mSv (360 mrem) annually.

Fig. 1.

the natural exposure and exceeds cosmic, terrestrial and internal as a source of exposure. Indeed the NCRP's pie diagram in 1987 (20) (Figure 1) demonstrated convincingly that radon was the major source of public exposure as had been emphasized earlier (25). Medical exposure of patients, partly because of the proper application of the effective dose equivalent for the partial exposures that mostly occur, is a smaller component than estimated earlier. The distribution of public exposures is similar in the U.K. and in many other countries (2) with only minor differences in the components.

Accidents. — An important development has been our experience with accidents. Before 1979, the accident rate was very modest (26), but TMI surprised many people even though it caused little or nothing in the way of radiological release or injury. It was followed however by Chernobyl and Goiania and the public has been increasingly nervous about the frequency of such accidents. Even so, the overall total of known deaths (101) for all accidents (305) for the years 1944 to 1989, when several million people, worldwide, are involved occupationally, is still small (27). The fatal accident rate has clearly increased in recent times, probably not unduly for a relatively safe modality and occupation, but increased nevertheless. What frequency can be expected in the future? Nobody knows! Professionals are more sanguine about accident frequency now than they were a decade ago however.

Nonionizing Radiation

Protection from electromagnetic nonionizing radiation and ultrasound has made very good progress especially as understanding of the nature of the effects has improved. There are good standards, partly due to NCRP (28), good measurement techniques (29) and a growing body of authoritative biological knowledge carefully evaluated by NCRP (28, 30) and published in NCRP reports 67 (31), 74 (32), 86 (33). It is evident that the evaluation of the effects of electromagnetic and ultrasound radiation, still not fully understood, need to be continued. Furthermore in this area too, better public understanding of the issues that concern them needs to be developed and fostered by NCRP.

Measurement and Dosimetry

In the measurement field, progress has moved us brilliantly from our one time inability even to measure background radiation well, to exquisite systems that can trace virtually every particle or photon and determine its energy and origin. Better systems are still needed for monitoring those exposed however (34).

In dosimetry, the dose is carefully specified now to organs and tissues, to the whole body, to cells and parts of cells and even to very small structures within cells by microdosimetric techniques. This is true for both external radiations and for most internally deposited radionuclides. On the macro-scale these properties are well known and can be relatively precisely measured or assessed (35). On the micro-scale progress has been made especially with the development of microdosimetry (36), but much more development and understanding is still needed before the precise sequence of events relevant to biological effects is fully documented.

Radiation Protection Philosophy

Our experience with standards in the radiation protection field, from an initial avoidance of acute effects to including in our protection selected susceptible population groups and controlling some natural sources, is a good demonstration of the success of human endeavors in this area, as pointed out in the opening lecture (37). Workers today get only a small fraction of the dose limits and public exposures are well controlled. Our problems tend to revolve around the relatively minor issues of exemption quantities and levels of little or no concern.

There are important problems in radiation protection control nevertheless, as Muntzing (38) points out. The United States does not speak with a single voice on radiation policies and regulations and it is no longer the leader in the international field in this area. Internally there are difficulties in understanding the scientific basis of radiation protection and in dealing with the political issues that emerge from them. These two should be separate but they demand an adequate level of understanding. The radiation literacy of the public needs improvement and it is time to emphasize scientific consensus rather than adversarial or litigious approaches. The radiological community has performed well but it should not be complacent. These problems, particularly those affecting the role of the United States in this area, need addressing.

The Future

Given all of the above, it could be argued that considering the progress that has been made in radiation protection in less than a century and especially in the last decade, it is not necessary to press further in improving protection of workers or the public, or even in understanding further the various processes by which radiation harms us. Some may think a stage has been reached in which enough is known and all that is needed now is a dotting of Is and crossing of Ts. But let us beware. There *have* been some surprises in the last decade. I can think of at least *three* fairly major matters (and probably many more minor) in the past decade alone. These three are, 1) the increase in risk estimates, probably about three times what we thought might be the case a decade ago, 2) mental retardation, previously known but now recognized as confined to a limited portion of the period of gestation during which the risk is estimated to be quite high, and 3) radon, now recognized to be a pervasive problem of public exposure, how serious has yet to be specified precisely but until there is evidence to the contrary it must be assumed that this is a general problem of rather significant proportions. Each of these three presents a new perspective in radiation effects from say, a decade ago, and must be taken into account in our appraisal of radiation protection progress and in our future thinking about controls, remedial actions and the like.

TABLE 4—*Needs for future radiation protection: Ionizing radiation*

- Radiation research, effects, mechanisms, epidemiology of risks.
- Sound application of principles, include ALARA, in all radiation work.
- Continue to improve measurement and dosimetry techniques.
- Public exposure problems, study and reduce radon levels.
- Accident prevention and preparedness.
- Recommendations and regulation to assure adequate control.
- Development of better public understanding of radiation as a hazard in context with other hazards.

One could easily add a fourth, 4) on the fact that we have now had at least two major accidents within the last decade. While perhaps these should have been expected, they were not.

One could easily add a fifth, 5) on public perception. Who among us would have thought, a decade ago, that public reaction to radiation would be so phobic. And that professionals would be virtually powerless, at least in the short term, to change it, but that is how it seems to be ... TMI and Chernobyl have probably been the most important factors in this, but fear of the unknown has a lot to do with it also.

One could go on ...

Future Needs in Radiation Protection and the Role of the NCRP

It is apparent that professionals need to continue to address radiation protection problems vigorously in order to maintain and improve the position of everyone with respect to radiation problems. It is not difficult to enumerate some of the major areas in which our efforts need to be placed and these are shown in Table 4 and Table 5.

What is the role of NCRP in all this? NCRP is directly involved in almost all of it. Indeed, NCRP can claim to be responsible for much of the knowledge that is brought to bear on radiation protection problems. The NCRP has taken a lead role in assessing the results of radiation research, both in basic work and in epidemiology. In this area, an anticipatory role is regarded as a special responsibility of the NCRP and one for which it is uniquely suited as long as its ties with the research

TABLE 5—*Needs for future radiation protection: Nonionizing radiation*

- Continue research on biological effects.
- Assess the hazards from electromagnetic and ultrasound radiations.
- Continue to develop sound measurement and control programs.
- Development of public understanding of the effects and the issues.

community stay close and firm. For example, NCRP has repeatedly drawn attention in recent years to the fact that risk estimates were edging upward and at the same time to the necessity to utilize the results of radiation research in establishing dose rate factors and quality factors. In this NCRP has performed a unique service and it must not only continue to do so but must redouble its efforts to ensure that the basic work needed in the laboratory and epidemiologically continues.

The NCRP has also spawned much of the work in application of good radiation protection practice, provided guidance, developed ALARA, etc. and been the primary source of professional knowledge. It continues to utilize the talents of health and medical physicists to further good practice and minimize exposures wherever possible.

—NCRP has taken the lead with public exposure problems such as radon, initiated the realization of radon's importance in 1983–84 by evaluating radon levels and risks (25, 39). NCRP has urged the further investigation of radon by EPA and DOE, and will continue to evaluate the risks and the levels and the means of reducing the exposure from the special vantage point of NCRP expertise. In addition all other sources of population exposure will be evaluated on a regular basis.

—In accident prevention and preparedness, NCRP has advisory committees that provide helpful advice and guidance for professional workers. If an accident happens however, the main *action* must come from the Government.

—Recommendations on basic exposure criteria are a special role of the NCRP. NCRP can provide what seems reasonable to professionals, based on their collective knowledge, recommended limits for operating personnel, for public exposure, for pregnancy, etc. Regulation however, its framing and its enforcement is the province of the Government. Hopefully, the Government finds NCRP guidance useful, and not competitive, which it is not intended to be, in their regulatory functions.

—In the development of better public understanding, regrettably none of us has done a good job, and "radiophobia" is still the dominant public reaction to radiation incidents. NCRP has been in the past, somewhat reluctant to engage in the field of public education. However the results of a recent questionnaire to the Council indicates that the Council members are now well attuned to this problem and want to see NCRP take the initiative. NCRP has several times in the recent past decade, offered public information programs to funding sources without much support from any of the recognized sources in our society. The problem is pressing, urgent and necessary and the NCRP is in a unique position to do it, because it can address the problem with more credibility than a government body. It is, however, a major undertaking, to establish a major public information office at NCRP and become the trusted informational resource the country needs. Furthermore it must be done, not at the expense of our scientific and professional program, but in addition to it. This task remains under consideration.

In all of the above the NCRP does not seek to compete with other interested parties who also have special roles, such as the various government agencies. The NCRP is a unique organization, it exists only to serve the public interest and no special interest of any kind. It perceives itself as a uniquely positioned partner of government in a society that badly needs the talents of the best informed.

I expect the NCRP will continue to serve the nation in the next decades to the best of the ability of its members and that its members will themselves always understand that this is the first obligation of the NCRP, to serve the "public" with scientific information and that the "public" in turn will, hopefully, appreciate the efforts the NCRP makes on their behalf.

References

1. Schull, W.J. "The status of somatic risk estimation," in *Radiation Protection Today—The NCRP at Sixty Years,* Proceedings of the 25th Annual Meeting of the NCRP (National Council on Radiation Protection and Measurements, Bethesda, Maryland) (1989).
2. UNSCEAR. United Nations Scientific Committee on the Effects of Atomic Radiation. *Sources, Effects and Risks of Ionizing Radiation,* Report to the General Assembly (United Nations, New York) (1988).
3. UNSCEAR. United Nations Scientific Committee on the Effects of Atomic Radiation. *Sources and Effects of Ionizing Radiation,* Report to the General Assembly (United Nations, New York) (1977).
4. BEIR. Committee on the Biological Effects of Ionizing Radiations. *The Effects on Populations of Exposure to Low Levels of Ionizing Radiation: 1980,* National Research Council (National Academy Press, Washington) (1980).
5. ICRP. International Commission on Radiological Protection. Publication 26, *Recommendations of the ICRP,* Annals of the ICRP Vol. 1 No. 3 (Pergamon Press, Oxford) (1977).
6. Neel, J.V. 1989 forthcoming Neel publ from RERF, 1989.
7. Abrahamson, S. "Genetic risk estimates: of mice and men," in *Radiation Protection Today—The NCRP at Sixty Years,* Proceedings of the 25th

Annual Meeting of the NCRP (National Council on Radiation Protection and Measurements, Bethesda, Maryland) (1989).
8. Otake, M. and Schull, W.J. "In utero exposure to a-bomb radiation and mental retardation: a reassessment," Brit. J. Radiol. **57**, 409–414, 1984.
9. Fry, R.J.M. "Experimental radiobiology and radiation protection studies," in *Radiation Protection Today—The NCRP at Sixty Years*, Proceedings of the 25th Annual Meeting of the NCRP (National Council on Radiation Protection and Measurements, Bethesda, Maryland) (1989).
10. NCRP. National Council on Radiation Protection and Measurements. *Influence of Dose and Its Distribution in Time on Dose-Response Relationships for Low-LET Radiations*, NCRP Report No. 64 (National Council on Radiation Protection and Measurements, Bethesda, Maryland) (1980).
11. UNSCEAR. United Nations Scientific Committee on the Effects of Atomic Radiation. *Genetic and Somatic Effects of Ionizing Radiation* (United Nations, New York) (1986).
12. Sinclair, W.K. "Experimental RBE's of high-LET radiations at low doses and the implications for quality factor assignment," J. Rad. Prot. Dos. **13** No. 1–4, 319–326 (1985).
13. ICRU. International Commission on Radiation Units and Measurements. *The Quality Factor in Radiation Protection*, ICRU Report No. 40 (International Commission on Radiation Units and Measurements, Bethesda, Maryland) (1986).
14. ICRP. International Commission on Radiological Protection. "Statement from the 1985 Paris meeting of the ICRP," in Annals of the ICRP **15**, No. 3, i–ii (1985).
15. NCRP. National Council on Radiation Protection and Measurements. *Recommendations on Limits for Exposure to Ionizing Radiation*, NCRP Report No. 91 (National Council on Radiation Protection and Measurements, Bethesda, Maryland) (1987).
16. Sinclair, W.K. "Quality factor, concepts and issues," in *Proceedings of the Tenth Symposium on Microdosimetry*, Rome, May 21–26, 1989, J. Rad. Prot. Dos.
17. Slovic, P. "Perception of radiation risks," in *Radiation Protection Today—The NCRP at Sixty Years*, Proceedings of the 25th Annual Meeting of the NCRP (National Council on Radiation Protection and Measurements, Bethesda, Maryland) (1989).
18. NCRP. National Council on Radiation Protection and Measurements. *Perceptions of Risk*, Proceedings of the Fifteenth Annual Meeting of the NCRP (National Council on Radiation Protection and Measurements, Bethesda, Maryland) (1980).
19. EPA. U.S. Environmental Protection Agency. *Occupational Exposure to Ionizing Radiation in the United States: A Comprehensive Summary for the Year 1980 and a Summary Overview for the Years 1960–1985*, EPA 520/1-84-005 (U.S. Government Printing Office, Washington) (1984).
20. NCRP. National Council on Radiation Protection and Measurements. *Ionizing Radiation Exposure of the Population in the United States*, NCRP

Report No. 93 (National Council on Radiation Protection and Measurements, Bethesda, Maryland) (1987).
21. NCRP. National Council on Radiation Protection and Measurements. *Exposure of the U.S. Population from Occupational Radiation.* NCRP Report No. 101 (National Council on Radiation Protection and Measurements, Bethesda, Maryland) (1989).
22. Baum, J.W. "Occupational exposures and practices in nuclear power plants," in *Radiation Protection Today—The NCRP at Sixty Years,* Proceedings of the 25th Annual Meeting of the NCRP (National Council on Radiation Protection and Measurements, Bethesda, Maryland) (1989).
23. Hendee, W.R. & Edwards, F.M. "Status and trends in radiation protection of medical workers," in *Radiation Protection Today—The NCRP at Sixty Years,* Proceedings of the 25th Annual Meeting of the NCRP (National Council on Radiation Protection and Measurements, Bethesda, Maryland) (1989).
24. NCRP. National Council on Radiation Protection and Measurements. *Natural Background Radiation in the United States,* NCRP Report No. 45 (National Council on Radiation Protection and Measurements, Bethesda, Maryland) (1975).
25. NCRP. National Council on Radiation Protection and Measurements. *Exposures from the Uranium Series with Emphasis on Radon and Its Daughters,* NCRP Report No. 77 (National Council on Radiation Protection and Measurements, Bethesda, Maryland) (1984).
26. *The Medical Basis of Radiation Accident Preparedness.* K.F. Hubner & S.A. Fry, Eds. (Elsevier/North-Holland, New York, Amsterdam, Oxford) (1980).
27. Lushbaugh, C.C., Fry, S.A., Sipe, A. & Ricks, R.C. "An historical perspective of human involvement in radiation accidents," in *Radiation Protection Today—The NCRP at Sixty Years,* Proceedings of the 25th Annual Meeting of the NCRP (National Council on Radiation Protection and Measurements, Bethesda, Maryland) (1989).
28. Wilkening, G.M. "Protection against nonionizing electromagnetic radiation: an evolutionary process," in *Radiation Protection Today—The NCRP at Sixty Years,* Proceedings of the 25th Annual Meeting of the NCRP (National Council on Radiation Protection and Measurements, Bethesda, Maryland) (1989).
29. Tell, R.A. "The state of the art in measuring electromagnetic fields for hazard assessment," in *Radiation Protection Today—The NCRP at Sixty Years,* Proceedings of the 25th Annual Meeting of the NCRP (National Council on Radiation Protection and Measurements, Bethesda, Maryland) (1989).
30. Carson, P.L. and Fowlkes, J.B. "Protection in medical ultrasound—modest progress in a 'low-risk' field," in *Radiation Protection Today—The NCRP at Sixty Years,* Proceedings of the 25th Annual Meeting of the NCRP (National Council on Radiation Protection and Measurements, Bethesda, Maryland) (1989).
31. NCRP. National Council on Radiation Protection and Measurements. *Radiofrequency Electromagnetic Fields—Properties, Quantities and Units,*

Biophysical Interaction, and Measurements, NCRP Report No. 67 (National Council on Radiation Protection and Measurements, Bethesda, Maryland) (1981).
32. NCRP. National Council on Radiation Protection and Measurements. *Biological Effects of Ultrasound: Mechanisms and Clinical Implications,* NCRP Report No. 74 (National Council on Radiation Protection and Measurements, Bethesda, Maryland) (1983).
33. NCRP. National Council on Radiation Protection and Measurements. *Biological Effects and Exposure Criteria for Radiofrequency Electromagnetic Fields,* NCRP Report No. 86 (National Council on Radiation Protection and Measurements, Bethesda, Maryland) (1986).
34. dePlanque, G., Lowder, W.N. and Beck, H.L. "Radiation protection measurement science: a long road traversed but a long way to go," in *Radiation Protection Today—The NCRP at Sixty Years,* Proceedings of the 25th Annual Meeting of the NCRP (National Council on Radiation Protection and Measurements, Bethesda, Maryland) (1989).
35. Kase, K.R. "Radiation dosimetry: past and present," in *Radiation Protection Today—The NCRP at Sixty Years,* Proceedings of the 25th Annual Meeting of the NCRP (National Council on Radiation Protection and Measurements, Bethesda, Maryland) (1989).
36. ICRU. International Commission on Radiation Units and Measurements. *Microdosimetry,* ICRU Report No. 36 (International Commission on Radiation Units and Measurements, Bethesda, Maryland) (1983).
37. Moeller, D.W. "History and perspective on the development of radiation protection standards" in *Radiation Protection Today—The NCRP at Sixty Years,* Proceedings of the 25th Annual Meeting of the NCRP (National Council on Radiation Protection and Measurements, Bethesda, Maryland) (1989).
38. Muntzing, M. "Development and current trends regarding regulation of radiation protection," in *Radiation Protection Today—The NCRP at Sixty Years,* Proceedings of the 25th Annual Meeting of the NCRP (National Council on Radiation Protection and Measurements, Bethesda, Maryland) (1989).
39. NCRP. National Council on Radiation Protection and Measurements. *Evaluation of Occupational and Environmental Exposures to Radon and Radon Daughters in the United States,* NCRP Report No. 78 (National Council on Radiation Protection and Measurements, Bethesda, Maryland) (1984).

Discussion

PAUL CARSON: Is there a concerted effort on internationally and nationally to make the best of what can be done with the Chernobyl accident in terms of human dosimetry that could still be done to measure the effects of the radiation?

WARREN SINCLAIR: Well, nationally of course, the Chernobyl accident, although a very serious one, has not impacted the American scene as you know, in quite the same way as it has in Europe and elsewhere. There is a very good review of the impact of the Chernobyl accident in the current 1988 UNSCEAR Report. Did you mean that we should capitalize on this experience to reach the public?

PAUL CARSON: No, I meant actually get a lot of pressure going to get in and make measurements on the highly exposed people in the 10 to 100 rad range.

WARREN SINCLAIR: Oh, I see, I'm sorry, I misunderstood you.

PAUL CARSON: With genetic type mutation measurements and so forth.

WARREN SINCLAIR: Of course, the people you're talking about are in the Soviet Union. They were exposed and there have been various assessments of the potential value of people exposed for epidemiological purposes, not only in the Soviet Union but in Europe as well. In general there's rather little that can be said about it in the case of Europe because the exposures that turned out are really not sufficient to provide a suitable opportunity. There is a unique population amongst those evacuated, and a few others, in the immediate Chernobyl area. There have been various representations from different international groups to engage the Russians, if you like, in a partnership about how to investigate those people. The Russians have been understandably somewhat onesided about it. They have assured us in meetings that they have the situation in hand. They have set up a new institute in Kiev for exactly that purpose. They are registering of the order of 200,000 people in that institute. One would expect numbers like 30,000 or so of those to be significantly exposed and perhaps contributors to realistic information. One can do calculations which show that a small group of the people, about 24,000 of them, could at least marginally be expected to demonstrate a small leukemia wave which could be starting about now. The Russians that I've been in contact with at least, assure us that they are well aware of all those problems. They have the registration of the people in hand, they are watching for leukemia and hopefully soon information will come from it.

TOR STROM, Livermore: I just want to mention Warren, that we at Livermore are presently negotiating with the Soviets through the IAEA to measure somatic mutations in some of these individuals that were exposed. We have made some measurements already, but we are now in the process of soon being able to do a broader study.

WARREN SINCLAIR: Yes, well thank you very much. I was aware of that. The Russians have been very interested in newer indices, if you like, of radiation exposure. They did and were offered by the IAEA immediately after the Chernobyl accident, the facilities of virtually the rest of the world through the agency, to do cytogenetic analyses, of all of the potential people who might be of interest. Unfortunately they didn't accept that at the time. It's a great pity because there was no way I think for them to have been able to carry out themselves the measurements that really would have been very helpful to all of us. Maybe, if it would have happened two or three years later in the new Glastnost and all of the rest of it, it might have been different, but right then they didn't accept it. Of course they were in something of a state of shock about the whole matter anyway and dealing with the acute victims was a major problem with them and they weren't looking ahead really at that point. It's a great pity they didn't accept the agency's offer.

SEYMOUR ABRAHAMSON: Warren, just to add another point. Both Dr. Patock and Dr. Fetkova, the director and vice-director of the institute at Kiev, just left Hiroshima about two weeks ago; and Dr. Fetkova will be spending four and a half months at RERF in Hiroshima and Nagasaki to look at the procedures that we're using both from computer banks standpoint of data collection and also to learn the techniques, some of which Dr. Turgis spoke about with respect to cytogenetic and cell mutation studies. So I think there will be a collaborative program set up between RERF and the Russian program.

WARREN SINCLAIR: Thank you, that's very good news.

RANDALL CASWELL: I would just like to comment that at the suggestion of Gil Beebe, the CIRRPC science panel is preparing a report on pre-disaster planning for human health effects research. And that report is, I suppose, maybe two or three months from coming out.

WARREN SINCLAIR: Thank you.

NCRP Publications

NCRP publications are distributed by the NCRP Publications' office. Information on prices and how to order may be obtained by directing an inquiry to:

> NCRP Publications
> 7910 Woodmont Ave., Suite 800
> Bethesda, Md 20814

The currently available publications are listed below.

Proceedings of the Annual Meeting

No.	Title
1	*Perceptions of Risk,* Proceedings of the Fifteenth Annual Meeting, Held on March 14–15, 1979 (Including Taylor Lecture No. 3) (1980)
2	*Quantitative Risk in Standards Setting,* Proceedings of the Sixteenth Annual Meeting, Held on April 2–3, 1980 (Including Taylor Lecture No. 4) (1981)
3	*Critical Issues in Setting Radiation Dose Limits,* Proceedings of the Seventeenth Annual Meeting, Held on April 8–9, 1981 (Including Taylor Lecture No. 5) (1982)
4	*Radiation Protection and New Medical Diagnostic Procedures,* Proceedings of the Eighteenth Annual Meeting, Held on April 6–7, 1982 (Including Taylor Lecture No. 6) (1983)
5	*Environmental Radioactivity,* Proceedings of the Nineteenth Annual Meeting, Held on April 6–7, 1983 (Including Taylor Lecture No. 7) (1984)
6	*Some Issues Important in Developing Basic Radiation Protection Recommendations,* Proceedings of the Twentieth Annual Meeting, Held on April 4–5, 1984 (Including Taylor Lecture No. 8) (1985)
7	*Radioactive Waste,* Proceedings of the Twenty-first Annual Meeting, Held on April 3–4, 1985 (Including Taylor Lecture No. 9) (1986)

8   *Nonionizing Electromagnetic Radiation and Ultrasound,* Proceedings of the Twenty-second Annual Meeting, Held on April 2–3, 1986 (Including Taylor Lecture No. 10) (1988)
9   *New Dosimetry at Hiroshima and Nagasaki and Its Implications for Risk Estimates,* Proceedings of the Twenty-third Annual Meeting, Held on April 5–6, 1987 (Including Taylor Lecture No. 11) (1988).
10 *Radon,* Proceedings of the Twenty-fourth Annual Meeting, Held on March 30–31, 1988 (Including Taylor Lecture No. 12) (1989).
11 *Radiation Protection Today—The NCRP at Sixty Years,* Proceedings of the Twenty-fifth Annual Meeting, Held on April 5–6, 1989 (Including Lecture No. 13) (1989).

Symposium Proceedings

The Control of Exposure of the Public to Ionizing Radiation in the Event of Accident or Attack, Proceedings of a Symposium held April 27–29, 1981 (1982)

Lauriston S. Taylor Lectures

No.	Title and Author
1	*The Squares of the Natural Numbers in Radiation Protection* by Herbert M. Parker (1977)
2	*Why be Quantitative About Radiation Risk Estimates?* by Sir Edward Pochin (1978)
3	*Radiation Protection—Concepts and Trade Offs* by Hymer L. Friedell (1979) [Available also in *Perceptions of Risk,* see above]
4	*From "Quantity of Radiation" and "Dose" to "Exposure" and "Absorbed Dose"—An Historical Review* by Harold O. Wyckoff (1980) [Available also in *Quantitative Risks in Standards Setting,* see above]
5	*How Well Can We Assess Genetic Risk? Not Very* by James F. Crow (1981) [Available also in *Critical Issues in Setting Radiation Dose Limits,* see above]
6	*Ethics, Trade-offs and Medical Radiation* by Eugene L. Saenger (1982) [Available also in *Radiation Protection and New Medical Diagnostic Approaches,* see above]

7	*The Human Environment-Past, Present and Future* by Merril Eisenbud (1983) [Available also in *Environmental Radioactivity,* see above]
8	*Limitation and Assessment in Radiation Protection* by Harald H. Rossi (1984) [Available also in *Some Issues Important in Developing Basic Radiation Protection Recommendations,* see above]
9	*Truth (and Beauty) in Radiation Measurement* by John H. Harley (1985) [Available also in *Radioactive Waste,* see above]
10	*Nonionizing Radiation Bioeffects: Cellular Properties and Interactions* by Herman P. Schwan (1986) [Available also in *Nonionizing Electromagnetic Radiations and Ultrasound,* see above]
11	*How to be Quantitative about Radiation Risk Estimates* by Seymour Jablon (1987) [Available also in *New Dosimetry at Hiroshima and Nagasaki and its Implications for Risk Estimates,* see above]
12	*How Safe is Safe Enough?* by Bo Lindell (1988) [Available also in *Radon,* See above]
13	*Radiobiology and Radiation Protection: The Past Century and Prospects for the Future* by Arthur C. Upton (1989) [Available also in *Radiation Protection Today— The NCRP at Sixty Years,* see above]

NCRP Commentaries

No.	Title
1	*Krypton-85 in the Atmosphere—With Specific Reference to the Public Health Significance of the Proposed Controlled Release at Three Mile Island* (1980)
2	*Preliminary Evaluation of Criteria for the Disposal of Transuranic Contaminated Waste* (1982)
3	*Screening Techniques for Determining Compliance with Environmental Standards* (1986), Rev. (1989)
4	*Guidelines for the Release of Waste Water from Nuclear Facilities with Special Reference to the Public Health Significance of the Proposed Release of Treated Waste Waters at Three Mile Island* (1987)
5	*Living Without Landfills* (1989)

NCRP Reports

No.	Title
8	*Control and Removal of Radioactive Contamination in Laboratories* (1951)
22	*Maximum Permissible Body Burdens and Maximum Permissible Concentrations of Radionuclides in Air and in Water for Occupational Exposure* (1959) [Includes Addendum 1 issued in August 1963]
23	*Measurement of Neutron Flux and Spectra for Physical and Biological Applications* (1960)
25	*Measurement of Absorbed Dose of Neutrons and Mixtures of Neutrons and Gamma Rays* (1961)
27	*Stopping Powers for Use with Cavity Chambers* (1961)
30	*Safe Handling of Radioactive Materials* (1964)
32	*Radiation Protection in Educational Institutions* (1966)
35	*Dental X-Ray Protection* (1970)
36	*Radiation Protection in Veterinary Medicine* (1970)
37	*Precautions in the Management of Patients Who Have Received Therapeutic Amounts of Radionuclides* (1970)
38	*Protection Against Neutron Radiation* (1971)
40	*Protection Against Radiation from Brachytherapy Sources* (1972)
41	*Specifications of Gamma-Ray Brachytherapy Sources* (1974)
42	*Radiological Factors Affecting Decision-Making in a Nuclear Attack* (1974)
44	*Krypton-85 in the Atmosphere—Accumulation, Biological Significance, and Control Technology* (1975)
46	*Alpha-Emitting Particles in Lungs* (1975)
47	*Tritium Measurement Techniques* (1976)
49	*Structural Shielding Design and Evaluation for Medical Use of X Rays and Gamma Rays of Energies Up to 10 MeV* (1976)
50	*Environmental Radiation Measurements* (1976)
51	*Radiation Protection Design Guidelines for 0.1–100 MeV Particle Accelerator Facilities* (1977)
52	*Cesium-137 from the Environment to Man: Metabolism and Dose* (1977)
53	*Review of NCRP Radiation Dose Limit for Embryo and Fetus in Occupationally Exposed Women* (1977)
54	*Medical Radiation Exposure of Pregnant and Potentially Pregnant Women* (1977)

55 *Protection of the Thyroid Gland in the Event of Releases of Radioiodine* (1977)
57 *Instrumentation and Monitoring Methods for Radiation Protection* (1978)
58 *A Handbook of Radioactivity Measurements Procedures, 2nd ed.* (1985)
59 *Operational Radiation Safety Program* (1978)
60 *Physical, Chemical, and Biological Properties of Radiocerium Relevant to Radiation Protection Guidelines* (1978)
61 *Radiation Safety Training Criteria for Industrial Radiography* (1978)
62 *Tritium in the Environment* (1979)
63 *Tritium and Other Radionuclide Labeled Organic Compounds Incorporated in Genetic Material* (1979)
64 *Influence of Dose and Its Distribution in Time on Dose-Response Relationships for Low-LET Radiations* (1980)
65 *Management of Persons Accidentally Contaminated with Radionuclides* (1980)
66 *Mammography* (1980)
67 *Radiofreqency Electromagnetic Fields—Properties, Quantities and Units, Biophysical Interaction, and Measurements* (1981)
68 *Radiation Protection in Pediatric Radiology* (1981)
69 *Dosimetry of X-Ray and Gamma-Ray Beams for Radiation Therapy in the Energy Range 10 keV to 50 MeV* (1981)
70 *Nuclear Medicine—Factors Influencing the Choice and Use of Radionuclides in Diagnosis and Therapy* (1982)
71 *Operational Radiation Safety—Training* (1983)
72 *Radiation Protection and Measurement for Low Voltage Neutron Generators* (1983)
73 *Protection in Nuclear Medicine and Ultrasound Diagnostic Procedures in Children* (1983)
74 *Biological Effects of Ultrasound: Mechanisms and Clinical Implications* (1983)
75 *Iodine-129: Evaluation of Releases from Nuclear Power Generation* (1983)
76 *Radiological Assessment: Predicting the Transport, Bioaccumulation, and Uptake by Man of Radionuclides Released to the Environment* (1984)
77 *Exposures from the Uranium Series with Emphasis on Radon and its Daughters* (1984)

78	*Evaluation of Occupational and Environmental Exposures to Radon and Radon Daughters in the United States* (1984)
79	*Neutron Contamination from Medical Electron Accelerators* (1984)
80	*Induction of Thyroid Cancer by Ionizing Radiation* (1985)
81	*Carbon-14 in the Environment* (1985)
82	*SI Units in Radiation Protection and Measurements* (1985)
83	*The Experimental Basis for Absorbed-Dose Calculations in Medical Uses of Radionuclides* (1985)
84	*General Concepts for the Dosimetry of Internally Deposited Radionuclides* (1985)
85	*Mammography—A User's Guide* (1986)
86	*Biological Effects and Exposure Criteria for Radiofrequency Electromagnetic Fields* (1986)
87	*Use of Bioassay Procedures for Assessment of Internal Radionuclide Deposition* (1987)
88	*Radiation Alarms and Access Control Systems* (1987)
89	*Genetic Effects of Internally Deposited Radionuclides* (1987)
90	*Neptunium: Radiation Protection Guidelines* (1987)
91	*Recommendations on Limits for Exposure to Ionizing Radiation* (1987)
92	*Public Radiation Exposure from Nuclear Power Generation in the United States* (1987)
93	*Ionizing Radiation Exposure of the Population of the United States* (1987)
94	*Exposure of the Population in the United States and Canada from Natural Background Radiation* (1987)
95	*Radiation Exposure of the U.S. Population from Consumer Products and Miscellaneous Sources* (1987)
96	*Comparative Carcinogenesis of Ionizing Radiation and Chemicals* (1989)
97	*Measurement of Radon and Radon Daughters in Air* (1988)
98	*Guidance on Radiation Received in Space Activities* (1989)
99	*Quality Assurance for Diagnostic Imaging Equipment* (1988)
100	*Exposure of the U.S. Population from Diagnostic Medical Radiation* (1989)
101	*Exposure of the U.S. Population From Occupational Radiation* (1989)
102	*Medical X-Ray, Electron Beam and Gamma-Ray Protection For Energies Up To 50 MeV (Equipment Design, Performance and Use)* (1989)

103   *Control of Radon in Houses* (1989)
105   *Radiation Protection for Medical and Allied Health Personnel* (1989)
106   *Limits of Exposure to "Hot Particles" on the skin* (1989)

Binders for NCRP Reports are available. Two sizes make it possible to collect into small binders the "old series" of reports (NCRP Reports Nos. 8–30) and into large binders the more recent publications (NCRP Reports Nos. 32–106). Each binder will accommodate from five to seven reports. The binders carry the identification "NCRP Reports" and come with label holders which permit the user to attach labels showing the reports contained in each binder.

The following bound sets of NCRP Reports are also available:

Volume I. NCRP Reports Nos. 8, 22
Volume II. NCRP Reports Nos. 23, 25, 27, 30
Volume III. NCRP Reports Nos. 32, 35, 36, 37
Volume IV. NCRP Reports Nos. 38, 40, 41
Volume V. NCRP Reports Nos. 42, 44, 46
Volume VI. NCRP Reports Nos. 47, 49, 50, 51
Volume VII. NCRP Reports Nos. 52, 53, 54, 55, 57
Volume VIII. NCRP Reports No. 58
Volume IX. NCRP Reports Nos. 59, 60, 61, 62, 63
Volume X. NCRP Reports Nos. 64, 65, 66, 67
Volume XI. NCRP Reports Nos. 68, 69, 70, 71, 72
Volume XII. NCRP Reports Nos. 73, 74, 75, 76
Volume XIII. NCRP Reports Nos. 77, 78, 79, 80
Volume XIV. NCRP Reports Nos. 81, 82, 83, 84, 85
Volume XV. NCRP Reports Nos. 86, 87, 88, 89
Volume XVI. NCRP Reports Nos. 90, 91, 92, 93
Volume XVII. NCRP Reports Nos. 94, 95, 96, 97

(Titles of the individual reports contained in each volume are given above).

The following NCRP Reports are now superseded and/or out of print:

No.	Title
1	*X-Ray Protection* (1931). [Superseded by NCRP Report No. 3]
2	*Radium Protection* (1934). [Superseded by NCRP Report No. 4]

3   *X-Ray Protection* (1936). [Superseded by NCRP Report No. 6]
4   *Radium Protection* (1938). [Superseded by NCRP Report No. 13]
5   *Safe Handling of Radioactive Luminous Compounds* (1941). [Out of Print]
6   *Medical X-Ray Protection Up to Two Million Volts* (1949). [Superseded by NCRP Report No. 18]
7   *Safe Handling of Radioactive Isotopes* (1949). [Superseded by NCRP Report No. 30]
9   *Recommendations for Waste Disposal of Phosphorus-32 and Iodine-131 for Medical Users* (1951). [Out of Print]
10 *Radiological Monitoring Methods and Instruments* (1952). [Superseded by NCRP Report No. 57]
11 *Maximum Permissible Amounts of Radioisotopes in the Human Body and Maximum Permissible Concentrations in Air and Water* (1953). [Superseded by NCRP Report No. 22]
12 *Recommendations for the Disposal of Carbon-14 Wastes* (1953). [Superseded by NCRP Report No. 81]
13 *Protection Against Radiations from Radium, Cobalt-60 and Cesium-137* (1954). [Superseded by NCRP Report No. 24]
14 *Protection Against Betatron—Synchrotron Radiations Up to 100 Million Electron Volts* (1954). [Superseded by NCRP Report No. 51]
15 *Safe Handling of Cadavers Containing Radioactive Isotopes* (1953). [Superseded by NCRP Report No. 21]
16 *Radioactive Waste Disposal in the Ocean* (1954). [Out of Print]
17 *Permissible Dose from External Sources of Ionizing Radiation* (1954) including *Maximum Permissible Exposure to Man, Addendum to National Bureau of Standards Handbook 59* (1958). [Superseded by NCRP Report No. 39]
18 *X-Ray Protection* (1955). [Superseded by NCRP Report No. 26]
19 *Regulation of Radiation Exposure by Legislative Means* (1955). [Out of Print]
20 *Protection Against Neutron Radiation Up to 30 Million Electron Volts* (1957). [Superseded by NCRP Report No. 38]

21   *Safe Handling of Bodies Containing Radioactive Isotopes* (1958). [Superseded by NCRP Report No. 37]
24   *Protection Against Radiations from Sealed Gamma Sources* (1960). [Superseded by NCRP Report Nos. 33, 34, and 40]
26   *Medical X-Ray Protection Up to Three Million Volts* (1961). [Superseded by NCRP Report Nos. 33, 34, 35, and 36]
28   *A Manual of Radioactivity Procedures* (1961). [Superseded by NCRP Report No. 58]
29   *Exposure to Radiation in an Emergency* (1962). [Superseded by NCRP Report No. 42]
31   *Shielding for High Energy Electron Accelerator Installations* (1964). [Superseded by NCRP Report No. 51]
33   *Medical X-Ray and Gamma-Ray Protection for Energies up to 10 MeV—Equipment Design and Use* (1968). [Superseded by NCRP Report No. 102]
34   *Medical X-Ray and Gamma-Ray Protection for Energies Up to 10 MeV—Structural Shielding Design and Evaluation* (1970). [Superseded by NCRP Report No. 49]
39   *Basic Radiation Protection Criteria* (1971). [Superseded by NCRP Report No. 91]
43   *Review of the Current State of Radiation Protection Philosophy* (1975). [Superseded by NCRP Report No. 91]
45   *Natural Background Radiation in the United States* (1975). [Superseded by NCRP Report No. 94]
48   *Radiation Protection for Medical and Allied Health Personnel* [Superseded by NCRP Report No. 105]
56   *Radiation Exposure from Consumer Products and Miscellaneous Sources* (1977). [Superseded by NCRP Report No. 95]
58   *A Handbook on Radioactivity Measurement Procedures.* [Superseded by NCRP Report No. 58, 2nd ed.]

Other Documents

The following documents of the NCRP were published outside of the NCRP Reports and Commentaries series:

"Blood Counts, Statement of the National Committee on Radiation Protection," Radiology 63, 428 (1954)
"Statements on Maximum Permissible Dose from Television Receivers and Maximum Permissible Dose to the Skin of the Whole

Body," Am. J. Roentgenol., Radium Ther. and Nucl. Med. 84, 152 (1960) and Radiology 75, 122 (1960)

Dose Effect Modifying Factors In Radiation Protection, Report of Subcommittee M-4 (Relative Biological Effectiveness) of the National Council on Radiation Protection and Measurements, Report BNL 50073 (T-471) (1967) Brookhaven National Laboratory (National Technical Information Service, Springfield, Virginia).

X-Ray Protection Standards for Home Television Receivers, Interim Statement of the National Council on Radiation Protection and Measurements (National Council on Radiation Protection and Measurements, Washington, 1968)

Specification of Units of Natural Uranium and Natural Thorium (National Council on Radiation Protection and Measurements, Washington, 1973)

NCRP Statement on Dose Limit for Neutrons (National Council on Radiation Protection and Measurements, Washington, 1980)

Control of Air Emissions of Radionuclides (National Council on Radiation Protection and Measurements, Bethesda, Maryland, 1984)

Copies of the statements published in journals may be consulted in libraries. A limited number of copies of the remaining documents listed above are available for distribution by NCRP Publications.